A BRIEF HISTORY OF MILK PRODUCTION

A BRIEF HISTORY OF MILK PRODUCTION

From Farm to Market

Bert Collacott

First published 2016

Published by
Old Pond Publishing,
An imprint of 5M Publishing Ltd,
Benchmark House,
8 Smithy Wood Drive,
Sheffield, S35 1QN, UK
Tel: +44 (0) 0114 246 4799
www.oldpond.com

A Catalogue record for this book is available from the British Library

ISBN 978-1-910456-52-1

All royalties from this book will be donated to Camsight. Camsight supports local people in Cambridge with sight loss.

The money they receive will be used towards their range of services for local people with sight loss. These include low vision equipment and technology centres; support for people in their homes and communities; rural social support groups in the villages around Cambridge; volunteer befriending; opportunities for taking part in sport and leisure activities including ten pin bowling, tandem cycling, swimming and tours of museums and galleries and groups and activities for visually impaired children and young people.

Book layout by
Servis Filmsetting Ltd, Stockport, Cheshire
Printed by Replika Press Pvt. Ltd, India
Photos by Bert Collacott unless otherwise indicated

Contents

Introduction

IN this book, the author – who was milking cows by hand at the age of ten years and has spent more than 40 years working with dairy farmers and milk processors – has devoted much time to describing the work of individuals and many organisations towards improving milk quality, the introduction of legislation, the setting of marketable standards, monetary incentives to dairy farmers to improve their product and harsh penalties for farmers producing milk not of legal standard, as well as rejection by processors, returning unacceptable milk to the producer. Early improvement in production methods on dairy farms is explained, such as on farm cooling of milk within 20 minutes of leaving the cow to storage in refrigerated stainless steel vats awaiting collection in modern insulated road tankers for delivery to urban processors for pasteurisation, and more recently microfiltration to provide top-quality unadulterated milk for consumers across the country.

For a number of years, a few people have chosen to condemn human consumption of milk and milk products. However, it continues to be an important staple food source. Nearly 50 per cent of milk produced in the United Kingdom is used for drinking and home cooking.

The part played by the coming of the railways in the mid-1800s is important, as it enabled raw milk to be transported quickly from distant farms to urban areas to meet the ever-increasing demand for fresh milk. The rise and fall of the wholly producer-owned organisation the Milk Marketing Board of England and Wales (MMB) is charted here. The MMB was wound up by the government after serving dairy farmers and the industry well for more than 60 years, which in the author's opinion contributed to the huge reduction in the number of farmers producing milk throughout England and Wales. The MMB's contribution to the dairy industry in improving milk quality, conversion of every dairy farm across England and Wales from handling of milk in churns to a more hygienic method of refrigerated storage and delivery to urban processors as well as orderly marketing of milk should never be underestimated.

1

Human Consumption of Milk From Bovine Cattle

ACCORDING to research at University College London, milk drinking started around 7,500 years ago in central Europe; the ability to digest the milk sugar lactose first evolving in dairy farming communities in this region. The genetic change that enabled these early Europeans to drink milk without getting sick has been mapped to dairy farmers who lived in a region between the central Balkans and central Europe.

Many reasons have been put forward as to why being able to drink fresh milk should be such an advantage. For example, milk can compensate for the lack of sunlight and synthesis of vitamin D in skin at more northern latitudes. Since vitamin D is required for calcium absorption, milk provides a good dietary source of both nutrients. Milk also provides a calorie- and protein-rich food source, which comes in a relatively constant supply compared to the boom-and-bust of seasonal crops; as well as being less contaminated than water supplies.

In its research, the university also goes on to highlight that traces of milk-related fats in archaeological relics pointed to dairying at the onset of farming in

England some 6,100 years ago. However, it is most likely that milk was first fermented to make yogurt, butter and cheese; not drunk fresh. The Romans used goat and sheep milk to produce cheese, and cattle as draught animals. However, Germanic and Celtic people practiced cattle dairying and drank fresh milk in significant amounts.

The current distribution of lactase persistence would seem to suggest an origin in Northwest Europe – especially Ireland and Scandinavia – since it is found at its highest frequency there today.

However, as the latest study suggests, dairy farmers carrying this gene variant probably originated in Central Europe and underwent more widespread and rapid population growth than non-dairying groups.

The spread of fresh milk drinking from the Balkans across Europe explains why most European lactase persistent people carry the same version of the gene, surfing on a wave of population expansion that followed the rapid co-evolution of milk tolerance and dairy farming. In Africa, there are four known lactase persistence gene variants and probably more yet to be discovered. Most are likely to originate from that continent, but the European version can also be found there too, especially among the Falani people. This diversity is probably the result of an 'imposition' of dairying culture on a pre-existing farming people rather than the natural spread of dairy farmers.

In recent years, there have been a few people who have chosen to condemn the human consumption of fresh milk and milk products. However, it continues to be an important staple food source.

In this book, the author draws on more than 40 years of experience from milking cows by hand while still attending school through to the early days of cattle-breeding by artificial insemination, as well as serving at various levels in the dairy industry across England and Wales.

The author also highlights the effect of the rise and fall of rail transportation of milk over long distances to urban 'liquid' markets (milk consumed daily in homes for drinking and cooking), as well as the orderly marketing of milk from farm to market destinations.

Other areas of interest covered within in the book are the change from churn handling of milk to the installation of on-farm refrigeration vats, enabling cooled milk to be collected in insulated road tankers for transportation over long distances to urban milk bottling plants; and the introduction of milk quality payment schemes that rewarded dairy farmers for a good quality product and penalised them where their milk was not of a certain quality standard.

2

Hard Times

U P until the advent of the railways in the mid-1800s dairy farming was very much a localised enterprise, serving farm workers and the local community with fresh milk and home-produced cheese and butter. In the main, milk was produced from local dual-purpose cattle, with surplus milk fed to calves for the beef trade or the odd pig or two to be killed and salted down for family consumption. Herd size was little more than five or six cows and production methods left much to be desired. Milking cows using a bucket and stool in fields was not uncommon.

Bedding was often a sparse covering of straw, even bracken from local commons or moorland in areas where straw was not readily available. Overnight, udder and teats invariably became contaminated with muck and were very likely not thoroughly washed before milking took place. Strip cups to detect mastitis were unheard of and filtering methods were not up to scratch, thus quality was poor.

In hot weather, souring often took place within 24 hours of production. Farmers started to realise the benefits of cooling milk as soon as possible after milking. Milk containers were sometimes immersed

Figure 2.1 Early rural delivery service.
From the collection of Norfolk writer and broadcaster Keith Skipper.

in a water trough overnight, or even in a brook that happened to be running through the farm premises.

To avoid the heat of the sun, dairy buildings were usually constructed on the north-facing side of the farm house. These dairies were well-ventilated through large openings protected by fine metal gauze to prevent flies getting access to milk stored in large open pans, as well as the cream, butter and cheese.

There were a variety of means of getting milk to the consumers. Where dairy farms were situated close to settlements, customers came to the farm to collect their milk. Small delivery milk rounds were done by milkmaids carrying four or five gallons of milk in containers often suspended from a wooden yoke placed across their shoulder; the milk being dispensed in pint or quart containers.

Figure 2.2 Early nineteenth–century milkmaid.

Population increase and rising living standards meant demand for milk became too much for on-foot delivery to householders, so dairy farmers started delivering using a pony and trap. They sold their milk from various size containers, usually 17-gallon tin-lined steel churns. Enterprising dairy farmers' wives made cheese and butter for sale. In the West Country, clotted cream, cheese and butter were produced and sold in pannier markets in nearby towns. This kind of dairying saw little or no change for many centuries.

Figure 2.3 Dairymaid churning butter for local markets, nineteenth century.

The making of clotted cream is quite a simple operation: milk is placed in a shallow pan, usually earthenware, and left in a cool room overnight for the cream to rise to the top. It is then scalded on a cooking range until clotting, left to cool again overnight, then the clotted cream is skimmed off.

Clotted cream has stood the test of time and is the main ingredient in Devon and Cornwall cream teas. It is also readily available through supermarket chains across the country. The resultant skimmed milk was used to fatten a pig or two for home consumption.

3

Town Dairies

DURING the 1700s and 1800s, the population of town and cities increased. For instance, in 1750 there were only two cities in England and Wales – London and Bristol – with a population of more than 50,000 inhabitants. By 1850, however, there was a total population of about 18 million, with 29 cities having a population of more than 50,000 and nine cities having 100,000 inhabitants. Families began to realise the value of milk and dairy products, living standards began to improve and demand for fresh milk increased.

Increasing population and improved living standards meant a greater demand for fresh milk in towns and cities, leading to transporting milk from far-off dairy farms by horse-drawn transport. However, this method was not suitable for transporting raw milk over greater distances to large cities and highly populated industrial towns, as by the time the milk reached the towns it would have often soured.

In the nineteenth century, the keeping of cows was not confined to rural areas. Many cows were kept in the centre of towns and cities such as Liverpool, Birmingham, Manchester and London. These were

Figure 3.1 Urban cow-keeper.

known as 'Town Dairies'. Owners of these establishments purchased fresh calved cows from country farmers.

The cows were milked to the end of their lactation and then sold on for beef. This is in sharp contrast to what happens today, where pedigree breeders record individual cow yields, retaining the best for breeding so that the farmer can be reasonably sure that their progeny will also provide good yields as well as constituent quality. This all helps to maintain a profitable enterprise.

Replacing dairy cows in Town Dairies became a lucrative market for those farmers who specialised in producing good-quality dairy cattle. Cows often came from far afield, there being accounts of them being driven to London by Welsh drovers who gained their

Figure 3.2 A cattle drover of the 1800s.

livelihood by selling milk produced by their cows en route.

The number of cows kept at Town Dairies varied from one or two secured by a chain around the neck in stalls behind a shop, to greater numbers kept in city open spaces. They were fed during the summer on fresh grass brought in daily, and in the winter on hay, brewers' grains, root crops and cabbage brought in by horse and waggon from outlying farms.

Many cow-keepers did not have very high hygienic standards of milk production; their cows often kept in insanitary conditions. Some were milked in the streets, which had open sewers running through them, and the milk was then carried in uncovered pails through these streets. Others were milked in their stalls behind the shops and the milk sold over the counter. Generally, cow housing was poor and not regularly cleansed. It was unlikely that cows' udders and teats were thoroughly washed and dried before milking began. Milking pails and the hands of those that milked the cows were also not properly washed.

Some people had a more enlightened attitude towards supplying raw milk to the public. In 1809, William Hartley opened a dairy in Glasgow and made sure that the cows and all utensils were thoroughly washed before milking took place.

In Holland Park, London, a herd of pedigree Shorthorns owned by Tisdale and Tunks enjoyed the luxury of 70 acres of pasture. The stalls in which cows were milked were glazed tiled and washed down daily. Improved premises and hygienic methods of production vastly improved the quality of milk supplied to consumers.

Until the middle of the nineteenth century, milk was sold in the streets of London by men and women carrying a shoulder yoke with a pair of wooden tubs. Each tub contained three or four gallons of milk. From the 1860s onwards, the increasing population and rising standards of living meant the demand for milk became too much for the yoke method of delivery so dairymen introduced low-wheeled carts, called 'hand-prams', which were pulled or pushed around the street by the rounds-man and his boy assistant. It was estimated that 'hand-pram' delivery persons walked around 15 miles each day. Later, as milk sales rose, horse-drawn floats were used. These could carry more milk and cover a greater area.

Because of poor keeping quality, there were three rounds each day. The first was immediately after early morning milking followed by a mid-morning round, named the 'pudding' round as the milk was hastily

Figure 3.3 Urban street-sellers.

made into puddings for the midday meal, and finally an evening round shortly after afternoon milking.

Milk was tipped from a wide-mouthed churn into a three-gallon hand-can, which was taken to the door. There it was ladled by a measured dipper into whatever container the customer provided.

If the customer was not at home when the milkman called, no milk was delivered. This led to the milkman providing half pint, pint and quart lidded cans that he would leave on the doorstep until he called again. Two problems arose with this system. Customers' containers were liable to be stolen from the doorsteps and it was more expensive for the dairymen.

Tipping milk from the churn into hand-cans also meant some customers got most of the cream which, being lighter, had risen to the top. The last customer received poor fat quality from the milk at the bottom of the churn. This unfairness was remedied by the introduction of delivery churns with a plunger and tap fitted to the bottom. The milk was then plunged from time to time during the delivery round. Local public health officials would, without warning, take spot check samples during rounds. Selling milk containing extraneous water usually led to dairymen being heavily fined.

Those frequently fined not only ran the risk of losing custom but earned the dubious title of the dairyman whose best cow was the water pump in the farmyard. The introduction of this law was concerned only with prevention of adulteration of milk and was not intended to improve the hygienic quality.

The town cow-keepers suffered badly as a result of outbreaks of cattle plague (rinderpest) that occurred between 1865 and 1866. The high milk-yielding Dutch Friesian, a popular breed introduced into this country, proved particularly susceptible to the disease. The outbreaks were so bad that the government introduced the Cattle Diseases Act of 1866, which ordered that infected cattle be slaughtered. This policy of slaughtering diseased animals led to the eradication of cattle-plague. It also virtually wiped out the stall-fed cows of London.

The difficulty in disposing of large quantities of dung from town dairies, which was causing a public health nuisance, created the necessity for the introduction of some of the first of many Public Health Acts.

A few town dairies survived up until the beginning of the Second World War. However, improved road and rail transportation of milk further contributed to their decline.

4

Railway Transportation of Raw Milk

THE development of railways in the mid- to late 1800s meant the opening up of the Midlands and London markets to dairy farmers in those remote areas where horse-drawn delivery of milk was impractical. The railways also offered a much cheaper form of transport of goods than that drawn by horse and cart. Seventeen-gallon milk churns were used to cope with the new railway transport. St Thomas's Hospital in London for many years had a contract with a nearby dairyman for supply of milk at one shilling per gallon. In 1846, they exchanged this for a supply from a Romford farmer a dozen miles away at ten pence per gallon, including transport on the Eastern Counties Railway.

In January 1865, the Great Western Railway Company moved a little under 9,000 gallons of milk to the London market. By January 1866, this figure had increased to 144,000 gallons.

By the 1880s, the main London rail stations, Paddington, Marylebone and Euston were becoming great milk trading centres.

In 1900, 25 million gallons of milk a year were being transported by the Great Western Railway. The

Figure 4.1 Milk arriving in 17-gallon churns at an urban railway station.

growth of the railways did not improve the condition of milk supplied to London and other large cities. Initially, it got worse.

Although there was a slight improvement in the compositional quality of the milk being produced because of better on-farm feeding and breeding policies, milk produced in the afternoons was not effectively cooled and stored for delivery to railway stations the following morning.

Added to the long railway journey, transporting by horse and cart to country railway stations did not help.

Yet despite the poor quality of the early railway milk, demand for it steadily increased, mainly because of its cheapness and availability.

5

Regulations to Improve Milk Quality

SADLY, some unscrupulous milk producers often made matters worse by adding water to their milk; in some cases impure water. Another alleged method of adulteration was to mix milk and snails' slime to make it froth and pass this off as cream. How people were convinced that it tasted like cream is hard to imagine.

Clearly not too soon, in conjunction with the 1855 Public Health Act, a permissive Act was passed creating Inspectors of Nuisances. One of their tasks was to ensure that 'any accumulation of dung, soil, filth or other obnoxious matter shall be removed by the person to whom it belonged'. But there were not many inspectors and not all were conscientious in their duties.

The first effective act, the Adulteration of Food, Drink and Drugs Act, was enshrined in law in 1872. This compelled local authorities to use their powers of inspection, with the back-up of chemical analysts, to aim to improve milk quality.

Unfortunately, in many of the more subtle cases where adulteration was suspected but not provable, so many people were becoming involved in the production and delivery of milk from the farmer to the consumer that there were occasions when the wrong person was fined. In 1875 a further Public Health Act made periodic sampling and inspection compulsory for all milk exposed for sale in England and Wales.

The Sale of Food and Drugs Act of the same year forbade the addition to milk of anything 'injurious to health'. But, as with so many of the Public Health Acts of that decade, they were not always comprehensive, so it was possible for unscrupulous farmers, wholesalers and retailers of milk to find loopholes.

The decision to prosecute offenders was left to the magistrates who were not expert bacteriologists or chemists. Even so, for the first time inspectors began to exert an influence on the condition of raw milk on offer to the public.

In 1885, regulations were issued under the Contagious Diseases Act of 1878, which required elementary cleanliness of those selling food. However, the first effective Acts did not come until the turn of the century. In 1899, the Sale of Food and Drugs Act was placed on the statute books. It gave the Ministry of Agriculture powers to regulate and fix minimum standards for the constituents of milk.

In 1900, the Wenloch Committee was established to consider the butterfat content of milk and cream. George Barham was a member of this committee, and his recommendations were embodied in the resulting Milk and Dairies Order (1901). This laid down a standard of a minimum 3.0 per cent butterfat and 8.5 per cent solids-not-fat, below which could be assumed to contain added water and deemed unfit for human consumption. But the law was concerned only with preventing adulteration.

Hygiene quality depended very much on smell or taste: more advanced and reliable testing was needed. The first such tests were for acidity. However, this did not necessarily take into account the 'natural acidity' of individual samples, which was known to be able to vary from area to area.

In 1893, Stokes Acidity Pellets, a combination of alkali and indicator, were available, which gave some idea of the amount of developed acidity in a sample if taken in conjunction with the natural acidity. This could be determined by direct titration and was the accepted method in some of the smaller creameries.

The Danish bacteriologists Professor Oria-Jenson and Bartell, introduced the Reductase test in the early 1900s. This was a dye reduction test using methylene blue as an indicator, which gave a fairly rapid guide to the activity of any bacteria that had contaminated a consignment of milk. It could be used in conjunction with the Fermentation Test, which was really a specific test for coliform organisms. These tests were not used much, if at all, in this country before the First World War.

Before the Ministry of Agriculture assumed responsibility for testing graded milk supplies, samples were taken by the local Sanitary Inspectors and tested at the Public Health laboratories. The Milk (Special Designation) Order of 1922 prescribed the colony count at 37°C and the coliform test as the official test.

The Milk (Special Designations) Order of 1936 changed the method of the examination of raw milk and the Ministry of Health Memo 139/Foods specified the four-and-a-half-hour and five-and-a-half-hour Methylene Blue test at 37°C in conjunction with the coliform test.

Figure 5.1 Continuous automatic bottling plant.
A.G. COLLACOTT COLLECTION.

Co-operative Societies became the largest wholesalers and retailers of milk in Britain and were pioneers of fast pasteurisation of milk sold in bottles on the doorstep and catering establishments, thus considerably improving the keeping quality.

It was not until 1880 when delivery of milk in bottles was introduced; the bottles were sterilised and reused. There was no change in bottle milk deliveries to customers until 1970 when a dairy in East Anglia introduced plastic sachets, this was quickly abandoned due to dogs and other animals stealing them off doorsteps. Today the majority of milk for the liquid market is sold in plastic containers.

The London Co-operative Society at the beginning of the First World War installed a milk sterilising plant, thus offering an alternative to pasteurised milk. This did not catch on, due mainly to the unpalatable taste.

By the 1920s, the Co-operative Society was retailing eight million pints of milk daily.

To safeguard and protect the consumer from possible disease, various Acts and Orders were passed. The Public Health Act of 1875 was passed to improve milk quality. There were three areas of legislation passed in addition to the Public Health Act, which affected the sale of milk. The first of these was the Sale of Food and Drugs Act 1875, which placed minimum composition standards on milk.

The Disease of Animals Act 1894 listed certain diseases that became notifiable. There was an increasing awareness of the causes and diagnosis of tuberculosis and possible dangers of drinking raw milk. The Production of Milk, the Dairies, Cowsheds and Milk Shops Order was passed in 1899.

Tuberculosis, a disease affecting both animals and humans, was a matter of considerable concern. In the 1930s, tuberculosis was present in approximately 40 per cent of dairy cows in the United Kingdom. At the time, the cattle were kept in fairly close confined conditions with especially poor ventilation, which allowed the spreading of the disease.

The tuberculosis lesions in udders allowed bacteria to enter the milk in the days before pasteurisation. This posed a great risk to humans drinking raw milk from infected animals. Some records show that more than 50,000 cases per year were occurring in the population. In 1935, the Attested Herd Scheme came into operation whereby owners of herds attested for bovine tuberculosis were paid 1d per gallon by the government for milk sold from their dairy herd.

The introduction of new regulations in 1949 hastened the acceptance of the need for greater care in the prevention of infection in milk and as producers improved their milking premises, a premium of 4d per gallon was paid by the government for milk from herds entirely free of tuberculosis. This proved a welcome opportunity for farmers to improve their income and invest in better housing for their animals.

Bovine animals are tested at regular intervals. Those found to be infected are slaughtered. This policy continues to the present time. The pasteurising of milk further reduces the risk to humans. A small quantity of untreated milk is still sold, with a warning to that effect. Herds supplying untreated milk are subject to strict and robust tuberculosis testing regimes.

However, tuberculosis in the nation's dairy herd remains high. There are also concerns that the disease is spreading into new districts where tuberculosis has not been for many years.

More than 30,000 infected animals were slaughtered in Great Britain in 2013, costing the taxpayer about £90 million. There appears to be some evidence that badgers are the cause, although this is disputed by those opposed to mass killing of badgers. Culling by shooting or gassing of badgers is a highly emotive issue, but there remains a huge challenge to our cattle industry. Hopefully the vaccination of cattle will be the answer. Clearly there appears to be no simple solution, but eradication of this awful disease must be the ultimate aim.

6

The 1920s

EARLY in the 1920s, farming went into decline, due mainly to the government's belief that importing food was more cost-effective. In some areas of the country, land prices fell dramatically. Owners of many large estates were offering farms to rent at record low prices and in some cases, to keep farms and buildings in good order, offered them as rent-free for a fixed period of time. The hitherto lucrative wool trade had long gone. Many Scottish farmers took advantage of low rents in East Anglia; some hiring entire trains to bring their farm equipment and animals, mostly Ayrshire dairy cows, to the nearest railway station and walking the animals to their new homes.

The Scottish farmer's high standard of milk production and the high cream content of milk from Ayrshire cattle soon found a ready market. (High cream line shown in bottled milk in those days was an excellent selling attraction to housewives.)

Specialist dairy farmers kept production costs as low as possible by feeding cattle home-grown fodder and sugar beet tops (crown and leaves), a by-product of the newly introduced growing of sugar beet, together with home grown hay root crops. Sugar beet pulp, a further by-product of sugar beet factories, became plentiful in wet form delivered to farms in bulk or dried in sacks for long-term or out-of-season storage. Due to extensive swelling when fed to cattle in dry form, dried beet pulp was soaked in water overnight before it could be safely fed to animals. A further by-product available to dairy farms situated near to breweries were brewers' grains. While this feed regime did not provide for high milk yields, it was reasonably cheap and kept cattle in good condition during the long winter months.

Specialist farming enterprises saw the beginning of the end of the traditional Norfolk four-course farming (wheat, turnips, barley, under sown clover or ryegrass for grazing) that kept farm soils in good order for decades.

The value of farmyard manure was much appreciated by growers, particularly on lighter land. It was quite common for cereal growers to provide straw bedding free of charge in return for farmyard manure. For many years up to the late 1940s, hundreds of steers for fattening in farmyards during the winter were brought into Norwich by rail from Ireland. This kind of enterprise made good use of abundant supply of straw, a by-product of extensive grain growing in the region, for bedding and production of valuable farmyard manure. Sadly this near 100 per cent organic farming has been replaced by artificial imported manures and chemicals.

Dairy farmers close to and supplying their milk to the liquid market with a good quality fresh product enjoyed a reasonably secure profitable market over other farm enterprises. Grain production, for instance, was suffering badly from cheap imports from America and Canada, making it more profitable to engage in milk, beef and pork production. An age-old saying among farmers in East Anglia was farming was either 'up horn, down corn' or vice versa.

While dairy farmers situated near urban areas were fairly well-off, those further afield were not so fortunate. In 1922, dairy farmers were paid eight pence per gallon by the major dairy companies for milk sold on the liquid market and five pence for milk sold on the lesser profitable manufacturing market.

Large urban milk processors, in addition to their own daily deliveries to householders, were supplying on a wholesale basis many small dairymen with bottled milk. While the liquid market grew at a slow and fairly constant pace, seasonal and daily fluctuations of quantities of milk coming off farms was a problem for those processors buying milk direct from farms. Processors endeavoured to solve this problem by offering dairy farmers a premium for a level daily quantity of milk. This proposal was not attractive to farmers who were anxious to sell their entire daily production.

To resolve this difficult situation, large urban liquid milk processors set up country depots; some with manufacturing facilities. Depots with no manufacturing facilities were usually situated in the east of England. Creameries engaged in manufacture, mainly cheeses and butter, were established in the less densely populated areas of southwest England and Wales.

Depots provided a valuable service for urban processors, taking in milk from local dairy farmers by lorry; cooling it before forwarding to the urban processor their daily requirement either by newly introduced glass-lined road tankers or similarly clad railway

Figure 6.1 1920s United Dairies glass–lined tanker used for transporting milk from country depots to urban processors.

tankers. Milk surplus to liquid requirement was sold off to the nearest manufacturing creamery.

As well as dairy farmers, milk manufacturers were also suffering the effects of cheap imports. They claimed that to make a profit they could not pay farmers more for milk used for cheese and butter production. This resulted in many dairy farmers being unable to cover the costs of production, let alone make their own profit. It was alleged that some manufacturers were paying farmers the manufacturing price for milk and then selling it on to the more profitable liquid market. Dairy farmers themselves were reported be undercutting one another.

There was no marketing organisation to protect dairy farmers' interests. Their position was woefully weak when negotiating with powerful dairy companies, especially the many thousands of farmers supplying the manufacturing market. Many gave up milk production; some near bankruptcy. The milk producing industry was in crisis, a situation that came back to haunt dairy farmers when they and the National Farmers Union did little or nothing to save their wholly owned Milk Marketing Board from being wound up by the government in 1994 in the ridiculous and mistaken belief that they would be better off taking on the all-too-powerful supermarkets and the dairy trade on their own.

The government recognised the severe problems that existed and tasked the Minister of Agriculture, Sir Arthur Griffith-Boscawen, to intervene. He took the chair at a meeting of interested parties. Thomas Baxter, a Staffordshire dairy farmer attended in the capacity of National Farmers Union member.

As a result of that meeting, the Permanent Joint Committee, including members from the trade, was set up and began discussions in October 1922.

What followed was in effect a quota arrangement, with each producer receiving a two-tier payment system that took into account seasonal production variation. For milk surplus to the liquid market, the dairy farmer would receive a lower manufacturing price that took into account the average quoted prices for Canadian and New Zealand cheese on the London Provision Exchange. There were no fundamental changes in the broad terms of this agreement until 1929 when the principal of basing a producer's quota on a seasonal variation was abandoned in favour of producers declaring their own standard quantity, incurring penalties if the 'declared quantity' were exceeded by an agreed percentage. Parity with imported cheese from Canada and New Zealand was also abandoned. It is worth remembering that these contracts were applicable only to certain buying areas of which London was the chief one, and that there was no binding condition upon producers and buyers entering into contracts. Moreover, there were ample opportunities for evasion on the part of both producers selling their milk at cut prices in the liquid market and buyers vending milk for liquid consumption for which they had paid the producer the lower manufacturing price.

There followed contracts less favourable to the producer. It became clear that unless some form of national cooperative marketing of milk was found, the industry would almost certainly collapse. Sir Arthur Street, a Dorset dairy farmer and NFU member, played an important role in drawing up the 1933 Agricultural Marketing Scheme. This provided a legal basis for cooperation of farm commodities if producers so wished.

In 1932, the Grigg Commission was required to examine the specific problems of the milk industry. Many of the ideas resulting from this research were reflected in a Milk Marketing Scheme drawn up by the National Farmers Union in 1933 based on the provision of the Agricultural Marketing Act.

As is often in a crisis situation, strong men come to the fore. In the case of the milk industry, this was none other than Thomas Baxter, a highly respected National Farmers Union council delegate, together with others such as John Dodd, a Cheshire milk producer who played an important role in the Clean Milk Campaign, and Sidney Wear of Yatton near Bristol who favoured a pooling system for payment to producers, which contained the gem of the machinery afterwards adopted for the Milk Marketing Scheme.

It was Wear at an Annual General Meeting of the National Farmers Union Council who put a case for a Milk Marketing Board in support of the following resolution:

> Having regard to the state of the milk producing industry of this country and lack of such efficient organisation of producers of milk which can ensure that the bulk of these producers will receive the price and benefits of any milk contract which the Permanent Joint Committee may enter into from time to time, and seeing this failure is generally brought about by a comparatively small number of these producers in each county, this annual general meeting of the National Farmers Union should approach the Minister of Agriculture and request it set up a Reorganisation Commission with the view to producing a scheme for milk in this country.

Wear's resolution was accepted by 97 votes to 94 with a third of the delegates present abstaining. R.R. Robbins and T.H. Ryland, both former presidents, resigned.

Once the decision had been taken, Thomas Baxter – whose early reservations had in any event been largely modified by the prospect of import controls offered by the 1933 Agricultural Marketing Scheme – threw himself wholeheartedly into preparation and promotion of the scheme.

The Reorganisation Commission was ordered by the minister on 18 April 1932 and its report was presented on 27 January 1933. A scheme largely based on the Grigg model was prepared immediately by the National Farmers Union and approved on 27 March 1933.

7

Milk Marketing Board of England and Wales 1933–1994 (MMB)

IT is beyond doubt that the MMB wholly producer-owned cooperative was the greatest commercial enterprise ever launched by milk producers, favouring both producers and consumers alike.

The dairy farming sector was in dire straits. Producers were in desperate need of a secure market for their milk and to be able to make a reasonable living. Under the Agriculture Marketing Acts of 1931 and 1933, the Milk Marketing Board of England and Wales was set up on 6 October 1933.

The marketing schemes required that the founding of a Board had to be brought to the attention of all persons likely to be affected and a poll of producers to be held. Sidney Wear of Yatton, who was chairman of the Somerset Milk Committee of the National Farmers Union for a number of years, put a resolution in 1932 to the National Farmers Union Annual General Meeting, which persuaded the union to support a Milk Marketing Scheme. I doubt if he envisaged that his resolution would provide a stable market for dairy farmers for the next 60 years.

The scheme was drafted by Mr Fairbairn of Ellis and Fairbairn who were the union's solicitors. Having

Figure 7.1 Meeting of the Milk Marketing Board, 6 October 1933 at Thames House, London.
A.G. COLLACOTT COLLECTION.

been drafted and approved by the National Farmers Union, it was presented to the government minister, Colonel Walter Elliot, and subsequently published in the *London Gazette*. After receiving objections a public inquiry was called for. Following discussions, the scheme was presented to milk producers of England and Wales for an initial poll. On 2 September 1933, more than 80 per cent of producers voted, of which 97 per cent were in favour of a wholly producer-owned Milk Marketing Board.

The provisional Board met on 1 June 1933 and appointed Thomas Baxter as its chairman. In 1934, Ben Hinds, a south Wales dairy farmer, was elected vice-chairman. The public inquiry, which lasted just over three weeks, began on 6 June. The scheme was approved by both Houses of Parliament on 27 July and the Board was constituted on 29 July. It met for the first time on 3 August 1933 and confirmed the appointment of Thomas Baxter as chairman.

Progressively, five Milk Boards were set up to serve

United Kingdom producers; the Milk Marketing Board for England and Wales (1933); Scottish MMB, the north of Scotland MMB and Aberdeen and District MMB (1933 and 1934); and Northern Ireland MMB (1955) under the Agriculture Marketing Acts of 1931 and 1933. The Milk Marketing Board of England and Wales began trading on 6 October 1933: the greatest landmark in our dairy industry.

The scheme was administered by a Central Board based at Thames House in London supported by 12 regional offices; ten in England and two in Wales. The Board was comprised of 12 elected milk producers, one for each region, together with three special members elected by producers and two co-opted members.

There were also Regional Committees comprising producers elected in each of the regions. To ensure a fair and balanced representation, each producer was allowed one vote for himself and one for each ten milk cows.

Not all dairy farmers were happy with the setting up of the Milk Board. Producer-retailers, of which there were more than 50,000, had a distinct advantage over wholesale producers, in particular those close to the more profitable urban liquid market as opposed to those supplying the less remunerative manufacturing market. Pool pricing and treating all producers equally, as well as the Board's disciplinary powers, did not appeal to all. Complaints from a minority of producers erupted in a noisy Second Annual General meeting in 1935. The chairman told the critics bluntly that if they thought they could do better without the Board, the remedy was entirely in their own hands. They could put the matter to the test by demanding a poll of registered producers to determine whether it should be wound up.

In July, a petition was received from South Wales' producers for the taking of a national poll on the issue. When the result was declared on 19 August 1935, it was certified that of the 98,458 dairy farmers voting, 81 per cent voted in favour of the continuation of the Board. This was the first and last attempt in 60 years by producers to wind up their own cooperative marketing organisation. The overwhelming majority in favour was further put to the test the following year when certain amendments to the scheme were tested by a poll and affirmed a majority of more than 87 per cent.

Like any other institution, the appointment of a general manager needed not only skill but an element of good fortune. The Board most certainly got this when they appointed Sidney Foster.

At the time, Foster was general manager of the London Co-operative Society. He had worked his way up from the bottom of the cooperative movement and was now in charge of a Society having a retail turnover of £10 million per annum, which included an annual throughput of 14 million gallons of milk. Foster went on to serve the Board, and England and Wales dairy farmers, for 15 years.

The purpose of the Milk Marketing Scheme in England and Wales was to bring stability to the beleaguered milk producing sector and to provide a meaningful rolling contract for milk producers by purchasing all the milk he/she wished to produce, marketing it in an orderly manner and paying the producer in the month following previous month's supply.

So as to have voting rights under the scheme, all milk producers in England and Wales had to register with the Board. Those wishing to sell their entire production to the Board entered into a General Wholesale Contract with the Board. Those wishing to sell their milk direct to consumers and catering establishments were granted a Producer Retail Licence by the Board.

Producer-retailers who wished to sell milk surplus to their retail sales were provided a Wholesale Contract in addition to their Retail Licence.

Special arrangements were made for farmhouse cheese-makers. This industry suffered a decline in the early 1900s. Farmers' wives and daughters, who were traditionally cheese-makers, were no longer content to work several hours a day in addition to their domestic duties.

During the 1920s, many farmhouse cheese-makers decided to sell their milk to large manufacturing dairy companies. The Milk Marketing Board endeavoured to remedy this trend when in 1934 the Farmhouse Cheese-makers Scheme was established, primarily at the request of cheese-makers themselves.

By 1938, there were more than 1,200 farmhouse cheese-makers using 24 million gallons of milk each year that would otherwise have been directed by the Board into the lower category market, thus reducing the price paid to all producers.

Sadly, World War II restrictions brought about a serious decline in the production of soft short-life-keeping varieties of cheeses and by 1954 only 126 farmhouse cheese-makers were registered with the Board.

With the winding up of the Milk Marketing Board in 1994, dairy farmers were soon to suffer a serious drop in the price they were paid for their milk and, finding themselves without the security of the Board, decided to look at other ways to save their dairy enterprise. Former farmhouse cheese-makers and a few other dairy farmers began making various soft cheeses for local markets and catering establishments. It is believed there are now more than 700 dairy farmers in England and Wales producing cheese on their farms.

The Board's first task in October 1933 was to secure contracts with 141,000 milk producers and 20,000 milk buyers and negotiate a £4 million bank loan to pay producers for milk purchased and sold by the Board in October in the third week of November. The Board's sole assets were a binding contract with producers and buyers whose only supplier was now the Milk Marketing Board of England and Wales. Milk buyers were contracted to pay the Board for all milk supplied to them by the tenth day of the month following supply.

In accordance with its obligation to treat all producers equally, all income from the sale of milk was placed into a pool (the Milk Fund) after deduction of administration and cost of transport of milk from farm to buyers' premises. For a time there was a small variation, depending on the producer's proximity to liquid markets. This was later abandoned.

It is perhaps worth noting that without the aid of present-day computer systems, 20,000 buyers and 141,000 statements were processed by hand and money due to producers was paid into their bank by 25 November. It was a remarkable achievement that the scheme got off the ground so efficiently and on time. In its first year of trading, the Board purchased

THE HAPPY PRODUCER.
"NOW AIN'T THAT LOOVELY, MY DEAR?"
(With Mr. Punch's congratulations to the Milk Marketing Board.)

Figure 7.2 An early cartoon celebrating the Milk Marketing Board.
COURTESY PUNCH.

and sold 3,890 million litres of milk. This was the beginning of one of the largest farmer-owned organisations that not only saved our milk-producing industry but went on to serve our dairy farmers well for more than 60 years.

Some producers were of the opinion that their own marketing cooperative was a government organisation staffed by a body of highly paid civil servants when, in fact, producers themselves were the sole owners with no middlemen. None of the Board members, all elected by dairy farmers, received a salary. Many of the staff in important positions had a farming background; in fact, many were farmers' sons. For the first time, milk producers had a secure lifetime market for all the milk they wished to produce at a fair price and were assured monthly payment. Bank managers recognised the value of the contract with the Board. Producers were able to secure bank loans for investment in their enterprises, even loans to purchase their farms and assist farmers' sons to get on the farming ladder.

The setting up of the Milk Board brought so much confidence and stability in milk production that by 1954, producers' numbers in England and Wales went up to 162,000 with an average herd size of 16 cows.

Up until the outbreak of the Second World War, the price of milk was subject to negotiation between the Board and the dairy trade. During the war years, the government became sole buyers of all milk from farms, thus the price to producers was controlled by them; the Board acting as an agent of the government, ensuring the producer was paid on time.

In 1954, the government handed back certain powers to the Milk Board. The price paid to farmers for milk sold on the liquid market and that paid by the consumer became part of the government's annual price review with the National Farmers Union. Milk sold for manufacture was subject to market forces.

The Milk Board and the dairy trade pursued a rigorous policy of improving the hygiene quality of milk. Early in the 1940s, a Resazurin dye reduction test was introduced to provide a 'sorting' test for milk delivered in churns to buyers' premises to detect contaminated samples or as a rejection test subsequently embraced in the milk buyers' contract with the MMB. It had the advantage of being quick and repeatable and it could be carried out by relatively unskilled labour with a minimum of laboratory glass or apparatus.

By 1963, all milk came from cows that had passed the tuberculin test (TT) and it was decided to alter the designation and transfer the responsibility for testing the quality of ex-farm milk to the 'dairy trade'. This was because part of the MMB Quality Schemes was designed to reward dairy farms on the quality of their milk and penalise those whose milk was not up to scratch. There were about 250 trade laboratories that operated the hygiene tests on behalf of the MMB. The test used was the more stringent two-hour dye reduction Resazurin test, which was known as the Hygiene Test, with price reductions for test failures. The MMB's Liaison Service was paid jointly by the MMB and the trade as a 'watchdog' to ensure uniformity in testing. This service was highly respected by both milk producers and the dairy trade. There were also tests for compositional quality (butterfat and solids-not-fat) as well as testing for antibiotic contamination. Milk contaminated with antibiotics caused serious problems in manufacturing of dairy products. Tests were also carried out for sediment as well as samples taken for monitoring of brucellosis cell counts and mastitis.

Once all milk was 'TT', the Milk (Special Designations) Regulations 1963 required that the designation 'Tuberculin Tested' for raw milk sold by retail should be replaced by the wording 'Untreated Milk'.

By 1979, when all milk was collected in bulk from dairy farms, it was felt by the MMB and the trade that the two-hour Resazurin test did not truly reflect the quality of milk and so the MMB decided to set up six central laboratories across England and Wales. These were in operation by October 1982. The new and sophisticated laboratory equipment provided more meaningful results over a much wider range of tests and could be backed up by automatic data processing. The weekly test on milk samples taken from farm vats by trained tanker operators were regularly monitored by MMB marketing officers, included fat, protein and lactose using the 'Milkoscan' machine and total bacterial count (TBC) for hygiene quality with the 'Petrifoss' machine. Other tests were for antibiotics, inhibitory substances and cell counts for monitoring mastitis. These tests provided the data for the quality payment schemes.

The trade laboratories continue to test for extraneous water and sediment. Where tanker-loads of milk that failed the antibiotics or extraneous water test could be traced back to an individual farm, the farmer was contractually bound to cover the cost of the entire tanker-load, less any recovered cash from a local pig farmer who would occasionally take the load for pig feed.

Local Authorities and Medical Officers of Health retained responsibilities for the pasteurised and sterilised milk regulations. Milk processors were obliged to keep daily records of pasteurisation and sterilising temperatures for local authority officers to inspect at regular intervals.

8

Changes on Farms

CONTINUING demand for milk meant changes on the farm. In 1889, William Murchland invented the first British continuous suction machine whereby the milk was drawn from the udder by sucking action. In 1895 it was followed by Dr Shield's 'pulsator' machine, which drew milk from the udder by a type of on/off sucking action, which was less painful to the cow than continuous sucking action. In 1902, Gillies developed a teat-cup that could be fitted more comfortably on to the udders.

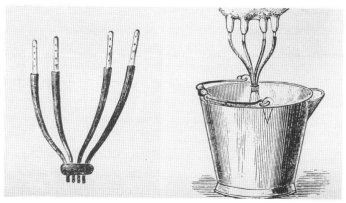

Figure 8.1 Early methods of machine milking.

These, and similar developments by the turn of the century, meant that the basic principal behind modern mechanical milking equipment had been established, thus substantially reducing labour costs on the farm.

But even by 1917, less than 1,500 herds in Britain were machine-milked. This, in the main, was because farmers complained that machine-milking reduced milk yields and lactation periods, as well as being expensive for small milk-producing enterprises. In 1919, a survey revealed that many farmers still preferred the more reliable hand-milking methods. Even the dramatic drop in milk prices and rapid rise in wages in the 1920s did not appear to have stimulated the introduction of cost-cutting machinery. In fact, as late as 1939, only 8 per cent of the herds (representing 15 per cent of the cows) in England and Wales were machine-milked and only those farmers with large herds seemed able to afford installing a milking machine and the running and maintenance costs. In fact well after World War II, many family farms were still milking by hand.

In 1922, Arthur Hosier, a Wiltshire dairy farmer, invented the Hosier Milking Bail. These bails, six stalls and three milking units mounted on skids, were ideal for outdoor milk production all year round, especially on free-draining chalk down land. The cow was secured in the stall by a chain round her hind quarters.

Each cow had its own feed hopper for feeding concentrates operated by lever by the cowman. The milk went direct to a sealed churn. At the end of milking, the cow was released by a rope operated by the cowman, which opened an exit-gate, allowing the cow to return to grazing with the rest of the herd.

Figure 8.2 Traditional hand-milking.
From the collection of Norfolk writer and broadcaster Keith Skipper.

These open bails were easily moved so as to prevent land becoming foul. Hampshire dairy farmer Rex Paterson quickly realised the economic advantages of movable bails and maximum use of grass, and established several herds on his free-draining land.

One man could milk and manage 80 cows producing around 50,000 gallons of milk each year. Hosier bails became popular among farmers who owned or rented grazing land some distance from the homesteads during the summer months. In East Anglia, open bails were used on summer rented marshes along river valleys. As herds became larger, static abreast and herringbone milking plants replaced open air milking. Mobile bails virtually disappeared with the introduction of bulk collection in the 1960s and 1970s.

Figure 8.3 Mobile bail direct to churn system – late 1940s and 1950s.
Courtesy Fullwood Ltd.

Figure 8.4 Cows were fed concentrated feed in accordance with its yield – 1950s and 1960s.
Courtesy Fullwood Ltd.

Figure 8.5 Fullwood Bucket Unit – cowshed installation 1950s.
COURTESY FULLWOOD LTD.

Figure 8.6 Fullwood eight-cow parladaptor – 1950s.
COURTESY FULLWOOD LTD.

Figure 8.7 Fullwood Abreast Parlour – 1960s.
Courtesy Fullwood Ltd.

Figure 8.8 Fullwood tandem parlour.
Courtesy Fullwood Ltd.

Figure 8.9 Fullwood 40-cow Index 90 Rapid Exit Parlour installed in purpose-built dairy unit – 1970s. Courtesy Fullwood Ltd.

Figure 8.10 Fullwood 40-cow direct to line, swing over parlour. Courtesy Fullwood Ltd.

Figure 8.11 Fullwood state-of-the-art 60 Rotary Abreast Parlour.
COURTESY FULLWOOD LTD.

Figure 8.12 Fullwood Merlin fully automated Robot Milker.
COURTESY FULLWOOD LTD.

Figure 8.13 Fullwood robotic milking at Pincots Farm.
COURTESY FULLWOOD LTD.

So how does robotic milking actually work? When robots are first installed on the farm, the cows are trained over two or three days to enter the robot. This is usually done by using it as a feed station to start with, enticing cows by adding their rations to the robot manger.

In most cases, training is very straightforward and within a few days, cows will naturally go to the robot by themselves to be milked.

Each cow has a unique system, this could be a neck or ear tag or a pedometer. These are recognised by the herd management system embedded in the robot.

When the cow enters the robot crate, the system will check whether she is due to be milked or not. If she is not ready, the restraining gate will open and she will be released until later. If she is ready to be milked the gate will remain closed and the pre-milking procedure of cleaning and stimulating the teats will begin.

The teat cluster is automatically connected to the udder as the system will have stored the location of the cow's teats, thereby ensuring the arm can move directly to the right place for safe, speedy and gentle attachment.

During milking, the ration of feed appropriate for each cow will be dispensed into the feed manger. Milking times will vary depending on the cow's yields; once milking is completed, the cluster will detach and then be thoroughly cleaned ready for the next cow to enter.

The latest Fullwood robotic milking machine – M^2erlin, includes a range of safety features and failsafes that not only improve cow welfare and operator safety but also protect milk integrity.

M^2erlin has to be strong enough to cope with cows standing on it, but it also needs to be gentle for milking. The M^2erlin arm has electric motors that can respond to tiny changes in current to give it a very soft touch. It is designed to be gentler than a calf's mouth – remember a calf has teeth and produces higher vacuum than a milking machine.

The first milking robots were developed in the early 1990s and the first commercial installations were sold in the UK in 1994; however, the technology has moved on incredibly in the past 20 years. The rate of uptake of the technology has increased dramatically in the past five years and now about 30 per cent of all milking systems sold in the UK are robotic.

Stringent hygiene testing at point of sale meant

improved methods of production on farms. In 1926, regulations permitted the use of sodium hypochlorite for sterilising of all milk vessels as well as inline cleansing of milking plants. Previously it had to be boiling water or steam chests. This method was time-consuming and costly to purchase and install. Milk buyers were obliged to provide clean churns and labels to dairy farmers supplying them.

To meet the growing demand for milk in the liquid market, producers progressively changed from breeding dual purpose cattle to high milk-producing breeds such as dairy Shorthorns, Ayrshire, Jersey and Guernsey cattle; the Channel Islands breeds popular for the high fat content of their milk.

Dutch Friesians had been known in England for several centuries. A government ban in 1892 forbade further imports because of outbreaks of pleuro-pneumonia on the Continent. The British Friesian Cattle Society was formed in 1914. It was followed by importations of highly selective stock from other countries such as South Africa and Canada. The breed proved to be especially popular because of its exceptional high milk yields.

Over the years, many pedigree Friesian herds were upgraded from our native Shorthorns. The red genes of Shorthorns occasionally produced red and white calves ('throwbacks' as they were commonly called).

Dairy farmers who moved away from mixed farming to specialised dairying quickly found ways of improving milk yields from cows specially bred for high milk yields.

At present, British and Holstein Friesian cattle make up around 90 per cent of the United Kingdom dairy herd. High butterfat milk from Channel Islands

Figure 8.14 Friesian dairy cow.

breeds such as Guernsey and Jersey is much favoured by dairy products manufacturers.

Dairy farmers for many years experienced great difficulty in obtaining fodder for their cattle during the winter months. Many were slaughtered and preserved in salt for family consumption and sale of meat in local markets. Improved conservation methods of fodder such as silage and hay meant high yielding cows could be saved and kept for breeding, thus improving production year on year. The use of clover and turnips in new rotation systems improved the quality of soil as well as raising the level of animal nutrition.

By the mid-twentieth century, many farmers had begun to see the value of feeding their cattle on mixed rations of pulped food with chaff, brewers' grain (a by-product of ale brewing), straw and hay, cabbage, kohlrabi and mangel-wurzel. Following the introduction of sugar beet growing in the UK in 1922, sugar

Figure 8.15 Channel Islands Jersey cows.
COURTESY RAWDONFOX.

beet tops and sugar beet pulp, a by-product of sugar production, were also a valued winter feed.

Forage crops were increasingly used towards the end of the century, along with feeds such as linseed, cottonseed and oilseed cakes. This type of winter feeding continued up to the 1950s.

Many farmers did not realise the huge benefits of improving grass as a means of fodder production. There was an ancient belief that pasture improved only with age and so fertilisers were not used on grass until the last few decades of the twentieth century, when farmers began to appreciate that grass did benefit from tender loving care.

Usage of well-rotted farmyard manure stored from in-lying cattle during the winter months was taken onto permanent grass lands by horse and tumbril, tipped in heaps and later spread by farm workers using a four-tine fork, followed by chain harrowing, as well as the use of chalk, marl and lime, depending on soil structure. Guano, a product of bird droppings and nitrate imported from Latin American countries was very popular, as well as potash from Germany. By 1900, there was also a growing awareness that each cow needed to be fed according to its individual output. As a result of these improvements, milk yields increased significantly.

Progressive dairy farmers quickly realised the benefits of keeping records of each cow's yield. National Milk Recording (NMR) was formed by the Milk Marketing Board of England and Wales (MMB) in 1943. Before then, milk recording was carried out by societies consisting of farmers in each county with some financial backing from the Ministry of Agriculture. The service provided a field officer calling at the farm at regular intervals to witness the weighing of each cow's yield as well as taking a sample of milk from each cow during evening and morning milking. Milk samples were tested at MMB laboratories for butterfat content. Both yield and butterfat content were recorded on each cow's data sheet. NMR was accepted by cattle breed societies for breeding data. The MMB provided about 30 per cent of the cost, the remainder was paid by the farmer.

NMR today is an integrated service provider working for both farmers and milk buyers as well as an independent source of data from advisors such as veterinary surgeons, farm consultants and breed societies. NMR is now totally independently financed with PLC status.

Milk recording data is used to provide the phenotypic database, providing information on traits in cattle offspring, behaviour and characteristic features, for UK genetic evaluation and is also used to provide the basis of food provenance schemes run by major retailers such as Tesco, Marks and Spencer and Sainsbury's.

NMR laboratories currently test up to ten million milk samples annually. This involves the processing of bulk milk samples from 97 per cent of UK farmers with every collection of milk from farms, and processing a monthly individual milk sample from 60 per cent of the cows in the UK. NMR is represented directly in England and Wales, Scotland and Northern Ireland as well as receiving samples from the Republic of Ireland.

Milk recording remains a voluntary service. However, the percentage of dairy herds that are recorded by the service increases each year as the professionalism required to manage a successful dairy enterprise on our farms, along with higher demands for food providence information as farmers, are viewed as the first step in the food chain.

As specialised enterprises evolved on farms, hitherto general farm workers' duties changed. Horsemen became tractor-drivers. Elderly workers finding it difficult to adapt to driving the new-fangled contraptions were occasionally heard to shout *whoa* instead of applying the brakes on tractors. General stockmen became cowmen/milkers and herd managers.

There was an increasing realisation of the need to improve the knowledge of farmers and their milk production workers and a number of education institutions were established. The first dairy school was founded at Nantwich, Cheshire in 1883, quickly followed by dairy learning establishments at Aylesbury and Reading.

Thus began the skilled training and research that was necessary to foster technological development and create modern scientific dairying.

9

A Cowman's Story – Arthur Thomas Godbold (1893–1968)

ONE must never forget those farmers and cowmen, rising between 4am and 5am, 365 days a year to milk cows, in many cases by hand, ready for collection by 8am. Arthur Godbold was one of the many rural young men across the UK who – more probably by accident due to high unemployment after the First World War than desire – became a cowman. He spent the greater part of his life as a cowman for one farming family, most of which was milking by hand. Arthur cared for the animals in his charge, never grumbled, and like so many others in the business of producing milk, he simply got on with the job even when his night was disturbed in helping a cow with a difficult birth of her calf.

Leaving school at 15 with above-average school achievements, Arthur did various odd jobs in and around his home in Rendham, near Saxmundham until he was enlisted into the army to serve throughout the First World War in the Dardanelles and Gallipoli. Except for malaria, which troubled him for many years, he returned home unscathed. Sadly, two of his brothers did not return. Again he had to rely on odd jobs for a living, primarily delivering provisions to country houses owned by the well-off members of the community. It was while on one of these visits that he met Gertie, his future wife.

Life took a turn for the better when in 1920 Arthur was offered the position of cowman by the Ashford family, who farmed in Rendham at one of their family member's farms at Gillingham, near Beccles. Gertie and Arthur were married and moved into a cottage in the village where they lived until his death in 1968.

Arthur took part in driving dairy cows from Rendham to Gillingham, approximately 30 miles. The animals, mainly Shorthorns, were the nucleus of a 50-cow dairy herd at Hill Farm, Gillingham, farmed by the Ashford family. Arthur was a junior member of three regular cowmen. Offer of promotion did not appeal to him. He was happy to remain a loyal under-cowman, rising at 4.30am during the summer months before heading to work rounding up the herd from Gillingham marshes for the start of milking at 5.30am. The method of production on the farm was of the highest order and milk from it in great demand from milk retailers in Beccles. The job was truly seven days a week. Time off was Wednesday afternoon each week. Around 1962, Arthur was awarded a long service medal by the Royal Norfolk Agriculture Society at their annual show – an award he was quietly proud of.

Determined to do his bit during the Second World War, Arthur joined the Observer Corps and became a plane-spotter at an observation post at Aldeby, passing information to defence control centres across East Anglia.

Arthur rarely mentioned his army service, except that he never ate bully beef again afterwards. However, he did achieve a little unsolicited fame in 1907 after he found an amazing bronze head of Claudius, Roman emperor and conqueror of Britain, in the River Alde in Rendham, near Saxmundham, Suffolk that, after his death, became the subject of the children's play *The Head in the Sand* by Julia Donaldson, Children's Poet Laureate 2011–2013 and author of the award-winning picture book *The Gruffalo*. The head is now owned by the British Museum.

Figure 9.1 Arthur Godbold and Tip.
A.G. Collacott collection.

Figure 9.2 Arthur Godbold (centre) and colleagues, pre-dehorning days and gradual introduction of Friesian breed into Hill Farm Shorthorn herd, 1932.
YVONNE FULLER.

Figure 9.3 Harvest at Hill Farm – sailing on the River Waveney 1950s.
COURTESY EASTERN DAILY PRESS.

10

Orderly Marketing of Milk as Operated by the Milk Marketing Board of England and Wales

IN 1942, the Milk Marketing Board took over full responsibility for collection of milk from farms. This made possible a considerable rationalisation of milk collection fields and a more orderly marketing of milk that substantially reduced transport costs for the producer.

This task, as well as maintaining it on a daily basis, was the responsibility of a team of marketing officers working across England and Wales. It was essential that these officers were fully conversant with the Milk Marketing Scheme, the Board's contract with whole-sale milk producers, as well as contracts between them and the Board, milk hauliers, raw milk buyers, and were able to foster and maintain good working relations with the dairy trade.

Orderly marketing of milk was the core operation of the Milk Marketing Board of England and Wales. The Board purchased each day all the milk a dairy farmer produced for sale and sold it to the most remunerative markets at the lowest marketing costs, to producers, processing dairies and manufacturing creameries.

Before a farmer could sell milk, he would be required to satisfy a Ministry of Agriculture Dairy Husbandry

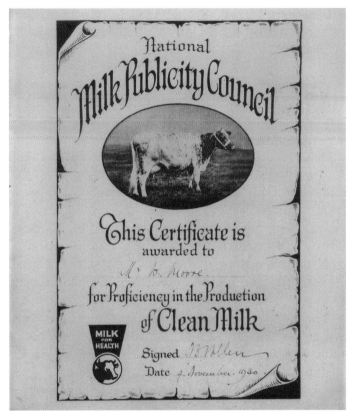

Figure 10.1 Clean Milk Certificate.
A.G. COLLACOTT COLLECTION.

Adviser (DHA) that his premises were of a standard to produce clean milk. Once registered, the DHA would advise the local Milk Board regional office.

The Board's area marketing officer would visit the farm and make himself known to the farmer. He would go on to explain the Board's wholesale milk contract or retail licence if the farmer wanted to sell milk direct to the consumer, together with his voting rights. If the producer's preference was a wholesale contract, the area marketing officer would wish to establish the proposed herd size so that he could market the new supply to the nearest liquid market requiring additional milk. Should the nearest processor be receiving adequate milk, then an alternative liquid market would be found by way of a 'buffering arrangement' through the nearest buyer's milk field. In simple terms, this involved directing a new supply in one end of the buyer's milk field and redirecting a similar size producer out at the other end. A buyer's contract did not specifically cover this kind of move, so good will was required all round.

A new producer may wish to produce Channel Islands milk from pure Channel Islands bred Jersey or Guernsey cattle. Channel Islands milk could command a handsome premium, provided a local buyer was willing to enter into a Channel Islands premium contract. There was in place a standard Channel Islands milk premium for producers agreed between the dairy trade and Quality Milk Ltd., but this could be varied upwards with the agreement of the buyer. Marketing officers needed to be very careful when marketing Channel Islands milk because the potential buyer may well be supplied by another buyer who in turn would lose trade. Officers always felt it best to be up front with buyers who in the main respected the officer's position. Give and take was always an important part of a marketing officer's job. It was not only producers who required the help of the Board's marketing officers but buyers as well.

The agreed Channel Islands premium was collected from the buyer by the Milk Board and the producer

Figure 10.2 Unloading full churns for onward collection, Devon 1952.
MUSEUM OF ENGLISH RURAL LIFE, UNIVERSITY OF READING.

was then paid in full; the premium in addition to the farm-gate price paid to all producers.

Once the formalities were dealt with, it was necessary to agree a collection point for this milk supply. In the days of churns, this would be a raised platform easily accessible for the collection vehicle. The usual height would be three feet six inches, and it would be of sufficient size to accommodate empty churns before transferring churns containing milk to the lorry.

Where a farm was inaccessible for collection vehicles, a roadside collection point would be agreed, mindful not to attract the wrath of the local Highways Authority. A six-figure map reference of the farmstead, as well as the collection point, would be recorded providing a valuable aid when negotiating payments to hauliers.

A first collection date would be agreed and the marketing officer would arrange for churns and labels to be delivered a few days before. The producer would be advised not to mix morning and afternoon milk and record this on the labels attached to each churn. In a few days, the farmer would receive from the Board's head office countersigned contracts, registration and formal notice of the haulier of his milk and the buyer to which his milk is directed.

At this stage, the farmer would be a member of a wholly producer-owned cooperative with voting rights in accordance with the Milk Marketing Scheme. A diligent marketing officer would call again in a few weeks to ensure the producer was happy with the arrangements made for him. He was also advised that his area marketing officer would from time to time arrange dairy farmers' meetings in his area to update producers on the Board's latest activities and to take questions on any relevant issues. These meetings provided an insight of the mood of producers, and for officers to explain that the remarketing of their supplies was necessary to meet market demand as well as to achieve savings in transport costs.

Each buyer had his own milk collection field of producers directed to him by the Board, some collecting from farms allocated to them using their own transport under contract to the Board and some by private hauliers also under contract to the Board as well as the Board's own transport; paid primarily on time and mileage. Transportation of milk from farm to buyers was paid out of the Milk Fund; therefore, a direct cost to the producer and an important area where marketing costs could be saved.

The Milk Board inherited a large number of small hauliers that collected milk from farms for delivery to small buyers. This mixed match of hauliers and various vehicle capacities made it difficult to rational-ise orderly transporting of milk from farms to buyers while at the same time providing a steady flow of milk into dairy premises. Nevertheless, with good will and the cooperation of all interested parties, some savings could often be achieved.

From time to time, officers would spend a day at the processor's premises, checking the time of arrival of loads of milk from farms and how well lorries and tankers were loaded, as well as turnaround times for second or third loads. These checks would provide an opportunity to rationalise a buyer's collection field and occasionally provide an opportunity to switch farmers on a temporary basis from a processor in surplus to another buyer in need of additional milk; or when a processor lost or gained additional trade. Over a period of time, these moves not only saved produc-ers money in farm collection but also payment to pro-cessors handling charges and secondary haulage costs. In addition to savings in milk collection miles, it was often possible during low milk production on farms to delicense collection vehicles for a short period. When organising these schemes, officers had to take into account fluctuating quantities of milk to be col-lected from each farm. For instance, a farmer could buy or sell a few extra cows. He was not obliged to inform the MMB of changes in production.

Importantly, when moving farms from one buyer to another and rationalising transport, officers needed to ensure buyers received a regular flow of milk coming into their premises. Nothing enraged a dairy manager more than seeing staff standing around waiting for milk to arrive as well as a processing plant not in use. Processors had milk delivery time arrangements with their bottle milk buyers and catering establishments. Drivers' working hours also had to be taken into consideration.

Rerouting farm collections occasionally brought about objections from producers whose collection times were changed, especially if to a much earlier time. In accordance with an agreement with the National Farmers Union, the Milk Board could not enforce a producer, without his agreement, to have his milk ready for collection before 8am.

One sharp-eyed officer found a small haulier routing one of his vehicles depending on when milk was available for collection, thus incurring more time and mileage than necessary. Money could be saved by re-routing the vehicle, but it meant three or four producers having their milk ready for collection much earlier than usual. A visit to each of these producers by the Board's local officer to provide an explanation for earlier collection went well. All were quite happy except one who said he had no intention of getting up at 4.30am to have his milk ready for collection at 8am and that the Milk Board would have to manage without his milk. It was necessary to explain to him that if his milk was not ready at 8am the following morning, it would not be collected, which would not be the Board's responsibility. Later that evening, the officer received a telephone call from the haulier saying he had received a call from the now very irate producer that his milk would not be ready and requested collection later in the day.

The officer firmly instructed the haulier to be at the farm at 8am. If no milk was ready for collection, he was to carry on his round and on no account return later in the day. At 7.45am the next morning, the officer was parked within sight of the producer's milk churn platform. By 8am the producer had placed two churns of milk on the stand ready for collection. The producer's herd size was four cows; not even enough for two MMB votes. This meant more money was saved at the risk of the Milk Board having to manage without his milk supply.

In agreement with the trade, each liquid milk buyer's daily requirement was reviewed annually by the Board. In the east of the country, March was found to be the month in which a buyer was closest to being in balance. To arrive at the buyer's daily requirement, the total quantity received from farms directed to them in that month was divided by 31.

It was then a simple matter of establishing the number of dairy farms producing the right quantity to be added, or removed, from the buyer's producer schedule.

This annual re-marketing exercise took place across the whole of England and Wales and was put into operation on 1 October each year. It was not without its problems in the east and southeast of the country. Milk production was in decline, which, coupled with an increasing population, meant extending the buyer's collection field further west.

For instance, the east and southeast of England liquid market requirement would need to come from the nearest least remunerative manufacturer creameries in the west of the country. Thus milk required, say, in Ipswich would be 'buffered' (buyers releasing farms on the eastern boundary of their collection field and replacing them with farms to the west) through several buyers' and hauliers' collection fields. Such moves often caused dairy managers to reschedule their processing programme and hauliers gaining or losing business.

Many producers found it difficult to understand why their milk was directed by the Board several miles east when their farm was situated a couple miles from a buyer to the west of their farm. This problem was overcome to a certain extent when churn collection was replaced with bulk collection from farms. This enabled milk to be reloaded into large tankers in collection fields for delivery over longer distances.

From time to time, milk hauliers' operations were surveyed in detail. Marketing officers would accompany a driver on his routes for the entire day. Not only time and mileage were recorded but other issues such as broken and unsafe churn platforms, too-small platforms and poor farm roads. Unreasonable hold-ups at farms or at a buyer's premises were noted and dealt with at a later date.

One officer, when carrying out one of these surveys on a small haulier, found the owner-driver delivering newspapers to the general public off what he considered to be the shortest route. The officer had no objection to delivering papers to farms from where he was collecting milk since it did not incur additional mileage or time, but most certainly not to the general public. As a result the officer gave the haulier a choice; paper round or milk collection. He choose the latter!

Another unusual survey was on a bulk tanker. This was in a bulk collection field where there was a very distant farm from the main cluster of producers. To survey this particular tanker, the officer had to be at the haulier's depot at 6am. The tanker arrived at the farm just before 7am to which, the officer assumed, the producer had arranged an earlier collection than that agreed with the National Farmers Union. The elderly producer and his wife of a similar age were busy milking. The officer asked the driver what the point was starting so early and arriving at the farm well before milking was completed. The driver asked the officer to follow him to the farmhouse. The breakfast table was set for four people. The officer was told that the fourth place was for him as the driver had told the producer he would be accompanying him on this particular day and, as the officer was staying overnight at a local hotel, he knew a hot breakfast would not be provided for him at 5am.

The driver was familiar with all that was necessary to prepare an excellent egg and bacon breakfast for four. Apparently he had entered into an agreement with the producer to prepare breakfast in exchange for early collection that allowed him to get to the main body of producers at 8am. Well, the driver deserved a break somewhere along the line!

On another occasion, when at a middle-sized processor's premises, an officer spotted four churns of milk on a lorry that also collected a supply from a large milk producer for delivery to a small buyer on his route. A tidy MMB authorised arrangement to market the small surplus to another buyer during peak production was in place. The officer checked with the lorry driver and found that the milk had not been unloaded at the small buyer's premises.

Further investigation found that the small buyer was claiming a handling allowance plus secondary transport from the Milk Board. This arrangement had been going on for a while. A sum of money was recovered and paid into the Milk Fund for the benefit of all producers.

11

The Marketplace

THE Board's selling price to buyers was a fixed price for milk sold on the liquid market as determined at the annual price review. Milk in excess of the liquid market was subject to various rebates depending on the final product, starting with a small allowance for high-category products such as fresh cream for domestic consumption, the catering trade and chocolate crumb to a much higher allowance for milk made into hard cheese, butter and skim milk powder, the latter mostly exported or sold back to farmers via animal feed merchants for calf rearing.

From the 1950s up until 1994, the MMB worked hard with the dairy trade to promote sales of liquid milk and milk for high category markets. Cookery demonstrations by qualified Milk Board staff to women's groups across the country helped sales of fresh cream and other dairy products. Milk Board mobile milk bars were prominent at agriculture shows and many other gatherings selling milk-based drinks as well as fresh milk.

Cool milk dispensers were installed in cafés, restaurants, MMB-owned milk bars on the high street, hospitals, places of work and other worthwhile venues.

For a number of years, the MMB sponsored the association football 'Milk Cup' as well as the 'Milk Cycle Race'. The Board's Sales Division arranged annual dairy queen competitions each autumn. These were very popular among dairy farmers' wives and daughters. So as to achieve maximum media and newspaper coverage, editors and local dignitaries were persuaded to judge these competitions.

Since the winding up of the Milk Boards, the liquid market has dropped from 60 per cent to 40 per cent.

Under the scheme, the Board were obliged to purchase all the milk the farmer wished to produce, plus producer-retailers' and cheese-makers' surplus milk.

Up to 31 July 1979, when churn collection from farms across England and Wales ceased, buyers were given a small allowance of around ½d per gallon to provide clean churns and labels to the producer. These were usually ten- or 12-gallon capacity and either made of steel with tin lining or aluminium. As a rough guide to volume, each churn had a raised mark on the inside from four gallons upwards.

During peak production periods and other variations in the daily quantity of milk coming off farms, buyers might receive more milk than they required. In this case, the buyer would report his surplus to the Board's regional office, who would then find the next nearest liquid market or, if this was not possible, the most remunerative manufacturing creamery. The secondary haulage was arranged and paid for by the Milk Board; the buyer paying the Board for all the milk he received from the farms directed to him, and in turn charging the processor receiving the surplus milk. Where milk was actually received at the first buyer's premises, relabelled and shipped out, a small handling allowance was paid by the Board. When a buyer was short of his requirement, it would be made up from the nearest liquid processor in surplus or from the nearest butter or cheese manufacturing creamery.

This arrangement meant milk churns belonging to various buyers finished up at another buyer's premises to which they did not belong. Each churn had embossed on them the owner's name. To overcome this problem, a company named Milk Vessel Recovery Services visited dairies and creameries on a regular basis collecting churns that did not belong to them and returning them, for a small fee, to their rightful owners.

In order to ensure daily delivery to the doorstep, it was essential that the liquid market be fully supplied every day of the year. To achieve this, the Board had highly skilled supplies officers on duty at Thames Ditton in Surrey and Newcastle-under-Lyme in Staffordshire to cover the whole of England and Wales for 365 days a year. Their busiest time was during the summer holiday season, when thousands of people went from inner cities and towns to seaside resorts, thus hugely increasing the additional milk requirements needed from dairymen supplying these resorts. At the same time, inner-city buyers were throwing up surpluses. Supplies officers would arrange, at short notice, tankers of milk to follow the holiday-makers.

The worst situation for supplies officers would be a sudden change in the weather causing large holiday cancellations. By that time, the additional milk would already be at the holiday resort buyer's premises; processed and bottled for delivery to hotels and boarding houses. This meant supplies officers making good on inner-city processors' unexpected requirements for returning holiday-makers and finding a market for the surplus milk thrown up by resort suppliers' buyers. Due to milk being a highly perishable commodity, these complicated marketing moves had to be done quickly.

It was in the producer's interest to ensure the Milk Board obtained the best possible price for all the milk that, at peak, came from as many as 160,000 producers across England and Wales. The most lucrative liquid market, which accounted for about 60 per cent of the milk coming off farms, had the benefit of a guaranteed price. The remainder went to manufacturing buyers who, in turn, were in competition with cheap imported dairy products; especially hard cheese and butter. These buyers also had to contend with seasonal variations in the volume of milk from farms directed to them, together with milk to and from the liquid market.

A secure market for all the milk a farmer wished to produce resulted in them increasing their herd size and investing in labour-saving equipment. This, together with the genetic improvement of dairy cattle through artificial insemination, brought about much more milk coming off our farms. Despite the Board's highly successful promotion of liquid market sales, by the late 1970s and early 1980s, more and more milk was being sold into manufacture at near or below farm production costs, especially at peak production periods of spring and early summer. This resulted in a reduced price paid to producers. Production was also increasing across Europe, which led to stockpiling of butter ('butter mountains' as they became known). Finally, in 1984 milk quotas were introduced.

12

New Milk and Dairies Regulations 1949

NEW Milk and Dairies Regulations introduced in 1949 gradually affected all milk producers, requiring their registration with the Ministry of Agriculture, which had been made responsible for the standard of dairy premises on farms.

Local authorities and Medical Officers of Health retained responsibilities for the Pasteurised and Sterilised Milk Regulations, for distributors of milk as well as administration of some of the precautionary measures if milk should become infected with disease communicable to humans.

Regulations laid down conditions for the use, construction and maintenance of buildings connected with milk production, handling, storage and processing of milk and cleansing of utensils.

Each county was to set up a Central Milk and Dairies Advisory Committee to enforce these new regulations to be administered by County Agriculture Executive Committees (CAECs). The procedure for a County Milk and Dairies Committee was as follows:

Figure 12.1 Surface cooler with 'D' bowl fed manually to allow milk to gravitate over external surface. Water, usually off mains supply, fed into hollow casing at bottom outflowing to waste at top. Popular with small producers from the 1940s up to introduction of refrigerated farm vats.

1. A County Committee shall meet from time to time but in any case not less than once in every six months.

2. A County Committee shall review the operation and administration of the Milk and Dairies Regulations and the Milk (Special Designation) (Raw Milk) Regulations in their county and may then report or make such recommendations as

Figure 12.2 Fullwood surface cooler for large herds receiving milk direct from round the cowshed pipeline milking or parlour systems. COURTESY FULLWOOD LTD.

they may think fit in respect of the operation and administration of these Regulations or any matter arising therefrom to the Central Committee.

3. A secretary shall be appointed by the Minister of Agriculture and Fisheries.

These Committees consisted of farmers together with representatives of various interests: for example, the dairy trade, usually a local dairy manager; the Medical Officer of Health; the Milk Marketing Board, usually the local area marketing officer; and the local Ministry Milk and Dairies Regulations Advisory Officer who would report conditions on a dairy farm that appeared on the agenda; none of whom had voting rights. The Milk Board officer would be expected to respond to questions concerning a producer's whole-sale contract or retail licence with the Milk Marketing Board or any other relevant issues. This was an excellent means whereby all the interested parties in the production of milk could gather and decide on the implementation of Milk and Dairies Regulations.

Matters that came forward for consideration by the committee were the registration of producers wishing to come into milk production, issue of designated licences, cancellation of registration and the suspension or revocation of designated licences for breaches of requirements.

Cancellation of registration was a very serious issue affecting a producers' livelihood, especially where milk was the sole income. It meant they could not sell any milk from their premises and the Milk Marketing Board ordered collection of milk from the farm to be stopped.

No registration was cancelled without a visit from two or three members of the Committee, together with a senior ministry official, to ensure that conditions were as reported to the committee by the DHA and for a report to be compiled and presented to the Executive Committee before a final decision was made.

One case revealed that the local Ministry Milk Officer, when carrying out a routine inspection of a farm, was dumped on a heap of farmyard manure by one of the producer's sons. There were no Committee volunteers for the customary farm visit. Only after much persuasion by the chairman did a couple of members agree to a visit.

Although there had been 'designated' milk produced for many years by some pioneers of the industry, the introduction of the new regulations in 1949 hastened the acceptance of the need for greater care in the prevention of infection in milk. Producers improved their milking premises and the government paid 4d per gallon premium for Tuberculin Tested milk.

The Milk (Special Designations) (Raw Milk) Regulations 1949 catered for three designations: Tuberculin Tested (Farm Bottled), Tuberculin Tested and Accredited. The aim was the complete eradication of tuberculosis from all cattle, especially the national dairy herd. Still not yet achieved, sadly the government's slaughter policy of infected animals still exists.

In addition to the approval of the methods of milk production, to obtain an accredited licence, dairy cows had to be clinically examined whereas for designation Tuberculin Tested cows had to pass the regular Tuberculin Test.

All these regulations and restrictions may appear to be unnecessary and tiresome. However, one has to take into consideration that milk is a highly perishable product that, if not produced in satisfactory buildings with high-standard methods, becomes contaminated with bacteria. While pasteurisation protects the consumer, milk needs to be in a good marketable condition before heat treatment. A good-quality product is a must in the marketplace.

13

Producer-Retailers

MANY producer-retailers in England and Wales did not welcome the setting up of the Milk Marketing Board. These made up nearly half of all producers. Selling direct to the consumers, they were better-off than wholesale producers and had a distinct advantage over processors and distributors. To overcome this advantage and maintain the principal of equality, producer-retailers were obliged to pay a small levy into the Milk Fund for distribution to all producers.

The Board's licence granted to producer-retailers meant accurate records of all their sales had to be sent to the Board each month together with the appropriate levy payment. Such records were subject to periodic audit by the Board's marketing officers.

It was important from the outset that producer-retailers were fully aware of the conditions of their retail licence, especially in regard to keeping records, and that their records would be audited by a Milk Board officer from time to time. Failure of compliance might result in the Milk Board taking disciplinary action with recovery of any unpaid levies and possible fines, plus an embarrassing report in the Board's monthly magazine showing the offender's name, size of fine and sum recovered.

Without such controls, producer-retailers could undermine the entire Milk Marketing Scheme. Under the scheme, producer-retailers were allowed to sell milk in containers up to one quart only and no more than ten gallons in bulk at any one time; for instance, to catering establishments.

Occasionally producer-retailers might wish to sell milk surplus to their retail requirements to the Board, especially during peak production periods. Under the scheme, the Board were obliged to purchase this milk in the same manner as milk from a wholesale producer. Sometimes a processor would be reluctant to accept producer-retailers' surplus milk especially if the producer-retailer was in direct competition on the doorstep with the processor. But because of a trend for producer-retailers to favour selling their milk to the secure market of the Board, the competing processor would wish to foster good relations with a competing producer-retailer in the hope of purchasing and absorbing that business into their own.

It was part of the Board's officers' duties to audit producer-retailer records. A few of these producers came up with various ingenious ways to avoid payment of this unpopular levy. Recordkeeping was sometimes poor, thus some hard bargaining would be necessary, usually with an adjustment of payment to the Milk Fund.

One producer-retailer whose records the Board's officer was tasked to audit was selling about 20 gallons of milk in his village. He could not resist the temptation to pull a fast one over the Board, or anyone else for that matter, and so, to his annoyance, officer audits were quite frequent. He was an elderly bachelor, a yeoman of the old school, salt of the earth, as the saying goes, and at best described as a likeable old rogue. He loved his cows, a couple of horses and according to local gossip one or two widowed ladies in the village. His retail milk records were almost non-existent. His monthly returns to the Board, always late, were little more than guesswork. The best clue the officer had was making a record of the number of milk crates, which he probably pinched, on his horse-drawn milk float and the number of cows he had on his small holding. He never failed to tell the officer how hard times were and how he was barely making a living, and how much better-off he would be working in the city, a likely square peg in a round hole.

For many years, his horse-drawn milk float caused much amusement in his village, especially to new-comers. It was a flat four-wheeled contraption with upright posts on each corner supporting a galvanised iron covering, allegedly to keep the milk cool during the summer months.

For a short time during the 1970s, the levy situation changed whereby the Board was obliged to pay producer-retailers a levy. The officer could not resist seeking him out on his milk round to see if times were still hard. 'No,' he said gleefully, 'trade is really picking up.' Time to call head office for a copy of his monthly returns, which were now coming in on time. Crafty old codger. How could one possibly dream of taking him to a disciplinary committee? It was not long after that he passed away.

In another instance, a complaint was made from a milk producer that his neighbour was supplying a large transport café 30 gallons of milk in churns each day. Such sales were known as 'off-contract' in breach of the producer's contract with the MMB and the Milk Marketing Act. The local officer was instructed to investigate, whereby he witnessed the delivery of three churns full of milk. When challenged, the producer became extremely abusive and refused any meaningful dialogue. The officer was left with no alternative but to prepare a case for the Board's disciplinary committee, who fined the producer a substantial sum.

However, these kind of cases were rare. Officers had to work within the Milk Marketing Scheme in the interest of all producers.

14

Short Measure

DURING the churn collection years, disagreements would occasionally arise between producers and milk buyers over the quantity of milk consigned by the producer. Consignments received at buyer's premises were tipped into a large stainless steel bowl and the quantity determined using scales that were regularly checked, tested and stamped by the local Weights and Measures officers. The scale operator recorded the total weight of each consignment and the producer was paid accordingly. Those whose consignments were found to be a half gallon or more below the total quantity shown on the producer's consignment labels attached to the churn were sent by post a shortage notice later that day or the next working day. Producers whose consignment exceeded the total on the label were credited the amount shown on the scales. Most producers determined the quantity using a raised mark embossed on the inner surface of churns at one gallon intervals from four gallons upwards.

There was, of course, the odd producer or two who would push their luck, hoping to get away with a gallon or two, especially when consigning milk on

Sundays or Bank Holidays, in the mistaken belief that their milk would not be weighed on those days. Occasionally a relief milker would overstate the consignment so as to show themselves in good light with the herd owner.

One producer, who employed a relief milker, returned from a long holiday and had to telephone his local officer to say he had difficulty in opening his door because of a vast heap of shortage notices from the dairy on the doormat.

A few producers genuinely considered they were being unfairly treated by their processor. The most common cause was that over the years, continuous stacking and denting of churn bottoms and sides made a mockery of the raised marks on the inner walls.

No producer liked receiving shortage notices and it did not bode well for producer–buyer relations. So, when at buyers' premises, officers would from time to time check the weighing of a few consignments, making a note of those that were short and if passing the farm later, the officer would drop in and tell the producer that he had witnessed the weighing of his milk and verified a shortage.

Another interesting case followed a telephone call from the local National Farmers Union secretary reporting a complaint from a producer that he was continually bombarded with shortage notices from the dairy receiving his milk. Following a discussion with the dairy manager, the secretary was invited, without prior notice, to visit the dairy and witnessed incoming milk supplies being weighed, the dairy manager adding that he did not like employing staff to spend time sending out shortage notices to dairy farmers.

In due time, a deputation of six milk producers, plus

the NFU secretary, turned up at the dairy without warning and watched their milk being weighed, only to witness three of the visiting producers' supplies were short. Apparently they did not stop to collect their shortage notices. I gather there were no further complaints.

Conversion from churns to bulk collection of milk did not completely eliminate measurement disputes. In the early days of bulk collection, milk was measured in a calibrated refrigerated farm vat by a trained tanker driver using a dipstick and then referring to a calibration chart to obtain the quantity. Occasionally a producer would query the tanker drivers' dipstick reading. This was nearly always due to the producer taking a dipstick reading before the milk was properly cooled. The compilation of the calibration chart was based on milk being at 4.5°C.

A bulk collection haulier reported to the Board's local office that a producer consigning around 4,000 litres of milk per day from two vats was having a running battle with his tanker driver over dipstick readings. He started to leave notes of readings he took before the milk was cooled as to what he should be credited, which was always to his advantage. The producer was rarely around when the tanker arrived but invariably put in an appearance when milk was being transferred from the second vat and would strongly dispute the driver's dip readings.

A few days later, the Board's local officer followed the tanker to the farm to find the producer's note for the tanker driver showing an incorrect advantage to him. The producer could not be found but made his usual appearance as the milk was being transferred to the tanker. The officer clearly explained that the correct reading and measurement was achieved when the milk was cooled to 4.5°C. The producer continued to argue his point and asked what his position would be if he refused the tanker driver's dip reading and demanded payment calculated from his dipstick reading. He was told that if this situation arose, the milk would be left on the farm and he would be given the option of a separate collection and the quantity determined over a public weighbridge, all at his cost. This must have been music to the tanker driver's ears because the following morning he left his tanker on the roadside, about 100 metres from the farm dairy. The producer spotting the driver walking to the farm dairy allegedly shouted to the driver telling him he had forgotten his tanker, to which the driver haughtily replied that he might not need it. The producer never again complained about milk measurement.

Another bulk producer came up with the novel idea of getting free milk. He carefully timed his run from the farm house to his dairy and plunged a gallon jug into the vat as soon as milk was being transferred to the milk tanker. The tanker driver on a number of occasions asked him to take his household requirements before the dip reading was taken, but the habit continued without the driver tanker raising the matter again.

Later, the producer remarked to the driver that he had not mentioned for quite a while about him taking milk after the dip reading was taken. The driver responded that he allowed a couple of gallons each day for his household requirements. The tanker driver insisted that he did not give a false dipstick reading but it certainly cured the problem, for it never happened again. If true, full marks to the tanker driver.

Processors were permitted to redirect tankers arriving at their premises to a public weighbridge without prior notice. These measures protected both producers and buyers. Weighbridge tickets were readily made available for examination to Milk Board marketing officers.

15

Further Incentives to Improve Ex-Farm Milk Quality

DURING the late 1950s, a body consisting of members of the Milk Marketing Board and representatives of the dairy trade set up what became known as the Joint Committee. Its aim was to improve the quality of milk coming off farms across England and Wales. It was agreed to introduce a self-balancing monetary incentive/penalty scheme for milk producers.

While the majority of buyers of ex-farm milk had their own quality control and milk testing facilities, the Joint Committee accepted that a scheme of such importance would require uniform testing across England and Wales. It was agreed that the already established Liaison Chemist service, a small team of chemists, would visit a buyer's laboratories to certificate existing milk analysts, as well as new analysts as they came on stream, as competent to test producers' milk supplies in accordance with the milk quality payment scheme. In order to ensure that all producers were included in the scheme, the Board engaged a small team of certificated analysts, equipped with mobile laboratories (Morris mini vans), to visit small buyers who did not employ analysts or have appropriate testing facilities to sample and test milk for the scheme.

A trial period of testing farm supplies showed that milk tested on one day in each calendar month provided a fair and representative standard on which to base constituent quality payments. In 1960, progressive quality payments were introduced. The first payment incentive was for butterfat and solids-not-fat content.

In 1982, total bacteria counts (TBC) were introduced, along with incentives to produce good hygienic quality milk and heavy penalties for those producers who did not.

The introduction of this test brought about considerable improvement in the hygiene quality of milk coming off farms in England and Wales. If manufacturers' instructions were not strictly followed when treating udders with antibiotics, then there was a possible risk of antibiotic-contaminated milk getting into the food chain as well as affecting manufacturing process. To combat this problem, heavy penalties were levied against those producers whose milk failed the antibiotic test and if proven, an individual producer might well be liable for any consequential losses incurred by a manufacturer. Such stringent measures brought about stricter control over the use of antibiotics in milk producing cattle. Some cooperative milk buyers provided, free of charge, facilities to enable producers to send in samples of milk from animals undergoing antibiotic treatment to ensure that the animal's milk was free of any traces of antibiotics before resuming the inclusion of her milk in the daily consignment.

It was also agreed with the Joint Committee that the Liaison Chemist service investigate test result disputes that occasionally arose between buyers' testing laboratories and producers, and arbitrate accordingly. Although there was provision for an appeal against the decision of the Liaison Chemists, the service was held in such high esteem that these were very rare.

Clearly incentive and penalty measures improved milk quality and the shelf-life of liquid milk delivered to the consumer. While most dairy farmers after World War II improved their milk production methods there was a hard core steeped in their forefathers' way of producing milk who had little or no interest in their product after it had left the farm.

16

Marketing Officers' Cooperation with the Dairy Trade to Improve Milk Quality

AS well as orderly marketing of milk and saving producers' transport costs, Milk Board area marketing officers worked hard with buyers' laboratory managers to improve the quality of milk coming off farms, visiting dairy farmers and advising them on how to maximise their income from their dairy enterprise, at the same time maintaining good relations with the dairy trade. Most producers welcomed advice to avoid their milk being rejected by their buyer. Sadly a few fell by the wayside either by their own accord or, in extreme cases, with the Board terminating their milk contract.

Throughout the 1960s and 1970s, many small processors sold out to larger dairy companies. This meant that in low production areas in the east of the country, milk in churns had to travel further to market, often well into the afternoon in hot weather conditions, thus making it more difficult to provide milk of good hygiene quality at the point of sale. Even so, many producers never had milk rejected on account of poor hygiene. During hot weather, officers would regularly visit processors' laboratories to examine producers' milk quality test results. Where appropriate, officers would visit the farmer whose milk was likely to fail the marketable test, telling the farmer his visit was not primarily a disciplinary issue but to prevent him working for nothing. Generally producers simply did not pay sufficient attention to washing and sterilising their milk equipment. Failure to replace perished teat cup liners, vacuum and milk tubes, and dirty claw pieces were also reasons for producing poor quality. Not cooling milk as quickly as possible after production, especially where milking took place in a mobile milking bail some distance from the farmstead, also had a detrimental effect on quality. Early morning visits often revealed poor production methods such as inadequate washing and drying of cow's udders before applying teat clusters. There was a knock-on effect of having to ship milk from distant country depots or manufacturing creameries to replace local rejected milk, adding further unnecessary cost of marketing.

One late-middle-aged producer – a very likable character with a nicely rounded figure and a similarly shaped, very red face, likely derived from his keen support of the local brewery over many years – was having his milk rejected on account of failing the routine hygiene tests. The Boards officer visited the farm on a hot summer afternoon and was directed by his wife, baby in arms, to a meadow where he was supposedly topping weeds with a tractor and cutter. The officer could not hear a tractor working but was attracted to loud snoring, which enabled him to trace the producer enjoying a nap in the hedgerow. Calling did not raise him from his slumber but a good kick to the sole of one of his boots did.

There was a strong smell of alcohol and he pleaded with the officer not tell his wife that he was found fast asleep. He proudly exclaimed that he was now a father of a beautiful little girl. The officer cunningly enquired as to whether the newborn was being reared on cow's milk. 'Cow's milk,' he replied, 'can't beat it.' Getting back to his dairy, the officer quickly dismantled most of his dirty milking equipment and suggested he would like to meet his wife and daughter. Off he went and brought them both to the dairy. After exchange of pleasantries the officer pointed out the reason for the poor hygiene quality of his milk.

The wife quickly got the message and began to contribute to the matter in hand. The officer made his excuses and left the farm, satisfied that the lady would adequately deal with the matter. Later that evening, the producer telephoned the officer to say that inspectors were bad enough to deal with, but the wife was worse. Apparently she sent him off to purchase new milk and vacuum pipes, thoroughly clean the milk plant sanitary bucket and vacuum lines. Naughty, but it did the trick. Some months later, the officer was told by a neighbouring milk producer that his visit had been the subject of discussion in the village pub, adding that he had never seen my roly-poly producer drunk, claiming after much consumption of ale and due to his shape, he simply fell over.

The Cambridgeshire Fens, well-known for growing fruit and vegetables for the London market as well as grain from the 1920s sugar beet, might well be considered an unlikely area for milk production. Up until the 1960s there were more than 60 small farmers engaged in dairying. In fact, the Fens had its own cheese-making factory producing the famous soft Cambridgeshire cheese produced mainly for local markets. This operation was moved to Bridge Farm

Dairies in Mildenhall, Suffolk in the 1970s but gradually declined and is no longer made in this country.

The majority of dairy herds in the Fens rarely exceeded 20 cows. They were devoted, in the main, to winter production when the farm's arable work was at its lowest and to provide an income for families through the winter months. Production from these herds was relatively cheap by way of feeding arable by-products such as substandard potatoes, cabbage and a variety of other root crops that were not considered fit for human consumption. Another benefit was an abundance of straw for bedding providing valuable farmyard manure for selected arable crops.

Man-made river banks necessary for the Fen drainage system and low-lying grass – aptly named the Washes due to flooding for a greater part of the winter – provided ideal cattle grazing during the summer. A few Fenland dairymen would dry their cows off and turn them on to the Washes during the summer, enabling them to concentrate on their arable enterprises. Others who produced milk all year round moved their summer production to the Washes using mobile milking bails, then bringing milk back to the farmstead for cooling and collection by MMB haulage contractors.

A secure milk contract with the Board to purchase all the milk they wished to sell enabled many young Fenland farmers' sons, as well as those interested in a farming career, to get on the farming ladder. Ken Robinson was a farmer's son; a fourth-generation Fenman. Ken was born in 1936 and left school at the age of 14 to help his father on his 120-acre mainly arable farm. By the age of 18, with the help of his father, Ken took the tenancy of 50 acres of grassland from Sutton Small Holders Association; a body set up to help young people to get into farming. Ken purchased his first farm animal, a heifer calf, for £1 10s from a dairy farmer about a mile from his home and brought it home in a wheelbarrow. It remained in Ken's ownership for its entire life, producing 19 calves; the first and last being the only pure whites she produced. Ken met and married Shcila in 1954 when they were both 18 years old. Their first decision was whether to purchase a badly needed cooker or their first cow. A cow was purchased for around £20, almost their entire savings. Gradually Ken and Sheila built up a small dairy herd that enabled them to begin selling milk to the MMB, thus providing a monthly income.

In the 1950s, it was possible to earn a reasonable living from 15 dairy cows. Ken and Sheila took over Ken's father's farm in 1966 and produced enough milk to install a 60-gallon farm vat. Ken remained in farming until his death at the age of 73 in 2009. His hobby was working with wood and much of his carvings and creations take pride of place in the family home. He loved his garden, which always seemed to be festooned with bright coloured flowers all year round. Ken was interned in his garden, just a few yards from the farmhouse. Ken and Sheila were a great couple. They produced four boys and four girls; all a great credit to them and none of whom caused their parents any trouble. Despite hard times, their children were always smartly presented and were regularly spotted walking in single file the mile and a half to the village school. Ken once reported to his local MMB officer that as he was a dairy farmer, his children did not benefit from the government's free milk scheme.

The officer explained that in such circumstances, the Milk Board acted as an agent of the government and arrangements were made to credit free school benefits on his monthly milk cheque. Later, the officer was told that at the time the benefit paid their house rent.

Ken carried on the family tradition of ice-skating on the Fens. No mean skater himself, he was immensely proud of his son Malcolm who excelled in the sport, winning many trophies in England and abroad. Ken and his family were true Fen folk. Ken was a generous family man who always had time for a chat with those he came in contact with. Head of a large family and dearly loved by Sheila, their children, grandchildren and great-grandchildren. How proud he would have been to read the following poem written by his 11-year-old great-granddaughter Beth at the time of his death.

My great grandad 'Robinson'
(By Beth)
My Great Grand-dad 'Robinson' worked hard upon
his farm,
His hips were sore, his legs no good, but still had his charm.
He sat next to the fireplace, rolling his cigarette,
With a grin upon his face which I shall not forget.
The things he said, they were so funny.
His Friday cake always looked so yummy.
He tossed his sweets across to us,
But never seemed to cause a fuss.
The stories he told, they made me laugh,
As he walked along the garden path.
The animals he had, there so many,
I remember best the Lab called Penny.
As he walked to view the chicks,
He used to wobble on his sticks.

And when he could no longer farm,
We found him in the big, brown barn.
There he was making things from wood,
All of these were incredibly good.

Today's the day we say goodbye,
Our hearts are sore and tears we cry.
He was my Great Grand-dad, I shall never forget,
If he's watching us, he will be laughing, I bet.

Figure 16.1 Kenneth Robinson.
COURTESY SHEILA ROBINSON.

17

From Churn to Bulk Collection of Milk from Farms

CHANGING from churn collection to collection of refrigerated milk in bulk from all farms producing milk for sale to the Board throughout England and Wales was a massive task. There were many issues to be addressed, not least resistance from the dairy trade, many of whom were reluctant to pay a premium of one penny per gallon for milk delivered in bulk, even though the hygiene quality of refrigerated milk was a good deal better.

Figure 17.1 1938 vintage farm collection vehicle.
COURTESY AUDREY AND SHIRLEY WOOLSTENHOLMES.

Figure 17.2 Milk collection by dedicated workers in all weathers.
A.G. COLLACOTT COLLECTION.

Many farm dairies required major alterations to accommodate farm vats that required a distance of two feet between the vat and dairy wall, some subject to local authority planning permission and all to comply with Milk and Dairies Regulations. A suitable electricity supply was necessary to power refrigeration units. Access roads and acceptable turnaround areas for tankers were essential. A concrete pad was necessary to ensure the collection tanker's hose did not become contaminated with dirt. Tenant farmers had to negotiate with their landlords where major construction works were required.

Occasionally it was necessary for marketing officers

to meet with County Highways officers to resolve issues such as visibility splays egressing onto the highway, together with weight restrictions.

Provision had to be made where access to tankers was not possible. Milk Board engineers produced a revolutionary design that enabled the cooling unit to remain at the farm dairy, while the milk container is taken to the roadside in the same way as churns. A simple trailer was provided for this purpose. Since both the milk container and the trailer are relatively lightweight, this equipment may be towed by either a car or tractor to an agreed level collection point (map referenced) to enable the mobile unit to be levelled for correct dipstick measurement using a jockey wheel.

Insulation of mobile vats was of paramount importance. Tests made at the National Institute of Research into Dairying showed that milk cooled to 40°F would rise in temperature by only 1.5°F over an eight-hour period when the surrounding air temperature is at 90°F.

Therefore, milk left at the roadside in a mobile container remained in excellent condition until collection took place, even on the hottest summer day.

The first mobiles to be produced had a capacity of 110 gallon and, later, 60-gallon capacity mobiles were available for the smaller producer. The refrigeration unit on the 110-gallon size was run by a small half-horsepower motor, so no difficulty was experienced with the farm's existing electricity supply. A small pump was attached to the vat enabling the tanker driver to rinse the inside of it to clear away any milk residue after transfer of milk to the road tanker.

The first on-farm refrigeration and bulk collection of milk in England and Wales was a joint innovation between Lord Rayleigh's Dairies at Hatfield Peverel near Chelmsford in Essex and Strutt and Parker Farms, begun in June 1954. Farm vats were imported from Wisconsin, USA. The milk was processed and bottled at Hatfield Peverel for next-day delivery to London hospitals and doorsteps in the city. In those days, housewives could be sure of getting fresh milk. Today it seems milk that is several days old and has travelled hundreds of miles is considered fresh.

The Milk Board set 1978 as a target date for all the milk coming off farms in tankers in England and Wales.

Despite reluctance by some buyers to pay a premium for refrigerated milk delivered to their premises in bulk, steady progress was made, mainly into Milk Board-owned processing and manufacturing plants. This progress was made possible by favourable premiums paid to producers depending on vat capacity and the Milk Board's vat rental scheme.

The vat rental scheme was of much benefit for those producers who were not in a position to provide capital outlay. 'Rent-a-Vat' was supported by an £850,000 EEC grant obtained by the Milk Board, as well as substantial large order discounts negotiated by the Milk Board with vat manufacturers. The EEC grant was worth 12 per cent of the vat purchase price and the manufacturers discount a further 20 per cent.

All farm vats had to meet the Milk Board's strict manufacturing and performance specifications and be capable of cooling milk to 4.5°C within 20 minutes of milk being discharged into it. Agitation timers were fitted to ensure milk was properly stirred before a sample was taken.

Each vat was fitted with adjustable legs, level points stamped and legs position sealed using special equipment by the Board's Area Marketing Officers. Officers also witnessed calibration of the vat, which required a calibrated vessel of known quantity approved by the Weights and Measures Authority being placed over the levelled vat and filled with water then discharged into the vat. When settled, a reading was taken from a dipstick peculiar to the vat and recorded. Water temperature was also taken into account in the preparation of a plasticised calibration chart at the Milk Board's head office.

One chart would be placed in the farm dairy to enable the tanker operator to convert dipstick reading to litres and a second copy placed with the milk buyer so as to cross-check the driver's conversion to litres.

Today tankers are fitted with meters that automatically record quantity as well as milk temperature, and a print-out is left at the farm.

Before the fitting of meters to tankers, tanker drivers were on occasion and without prior warning, diverted to a public weighbridge for weighing before discharging their loads. This assured good quantity control for both producers and buyers. Milk tanker drivers underwent rigorous training to examine and stir the milk well before taking a sample in the correct manner, measuring the amount of milk in the farm vat, recording the milk temperature and issuing a receipt. Their performance was regularly checked by the Board's marketing officers. Initially, some milk buyers were reluctant to hand over this responsibility to tanker drivers, even though they were able to reject tankerloads of milk failing the contracted marketable test and the right to weigh loads over a public weighbridge. In fact, there was little or no overall difference between milk delivered in churns and bulk intake.

It was decided that bulk development should commence in regions where the number of churn producers was relatively small.

The programme started in October 1973 in four regions in the south and east of England, followed by the West Midlands, northwest, far north, southwest and Wales. All farms supplying churn milk in these regions were visited by bulk development advisers, giving producers the opportunity to discuss individual problems they may have in converting to bulk collection.

Before the advisory visit, an initial survey was made by the Board's area marketing officers to discuss a producer's future production plans. It was vital that the herd size be projected for five years and that the vat they intended installing would be large enough for future production.

It was important to make clear to producers that any milk surplus to vat capacity in a 24-hour period would not be collected.

After many years of stalling, all buyers agreed to accept ex-farm milk in bulk. Farm vat and milk tanker manufacturers barely kept pace with demand. Buyers incurred significant costs in installing greater storage as well as fast discharge pumps to Milk Board specification to facilitate quick tanker turnaround (twenty minutes) and tanker in-place sterilising systems for farm collection tankers at the end of each day. It was important that area marketing officers worked with buyers and hauliers to ensure a constant flow of milk into their premises. Equally, it was essential to ensure tankers were not held up waiting to be unloaded, thus making them late and clashing with afternoon milking on farms. Achieving a constant flow of milk and maintaining farm collection vehicles as full to capacity as possible to processors during the transitional period of changing from churn to bulk handling of milk, especially where for a time milk was delivered both in churns and bulk proved difficult.

Most area marketing officers had a number of buyers in their respective areas. Fortunately, as bulk collection neared 100 per cent, a satisfactory flow of milk into buyers' premises quickly settled. Marketing officers were required to ensure that milk collection was run efficiently all year round. The aim was not less than 85 per cent of tanker capacity.

In 1974, more than 4,500 farm vats were installed and churn collection finally ceased on 31 July 1979.

By this time, the number of producers in England and Wales had dropped to around 43,000 with an average herd size of just under 60. National production remained much the same.

Early farm collection tankers had a carrying capacity of 1,750 gallons (8,000 litres) and were fitted with petrol driver pumps to transfer milk from farm vat to the tanker. These were soon replaced with 2,000-gallon (9,000-litre) capacity tankers with vacuum loading and discharge systems. Most farm collection tankers were single compartment, although a few had twin compartments in order to keep Channel Islands premium milk separate. All tankers were manufactured to strict Milk Board specifications. The carrying capacity of each tanker was more than twice the capacity of churn collection vehicles. Thus considerable savings on haulage cost were achieved and enabled milk to be hauled over much longer distances.

The benefits of vacuum filling and discharge systems on farm collection tankers (primaries) enabled milk to be discharged without exposure to the elements to 4,500 gallons (20,000 litres) and reloaded in tankers in the collection field at pre-arranged meeting points, country lay-bys, large farmyards or established tran-shipment depots for delivery to distant buyers.

Tanker trailer units were introduced, increasing the capacity to 4,000 gallons (18,000 litres). These units were adapted so that the trailer could be left at a stra-tegic point in the milk collection field and fed by the towing tanker or perhaps two primary tankers operat-ing the same area. The introduction of this system saved transport costs, especially in areas where pro-ducer numbers were low and collection fields a long way from buyers' premises.

Bulk handling of refrigerated milk from farms over long distances around the clock 365 days a year was all part of orderly marketing; not only haulage savings for

Figure 17.3 300–gallon refrigerated farm vat, circa 1960.
Bill Cornwall collection.

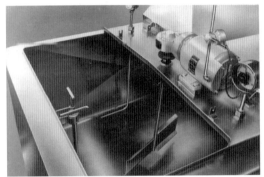

Figure 17.4 Farm vat showing dipstick, thermostat and agitator paddle.
BILL CORNWALL COLLECTION.

Figure 17.5 Tanker operative inspecting milk before vacuum transfer to road collection tanker.
BILL CORNWALL COLLECTION.

producers but the cost of handling, cooling and transhipping milk through country depots. All country depots were closed following completion of on-farm refrigeration.

All dairy farmers whose milk is collected in bulk from farm vats need to take into consideration the possibility that heavy snow, floods or other adverse conditions could prevent the tanker from reaching the farm at its usual time. Other circumstances, such as industrial disputes, could delay the tanker's arrival by several hours.

Figure 17.6 Tanker operative reading dipstick graduation mark.
BILL CORNWALL COLLECTION.

Figure 17.7 Mobile vat designed by MMB engineers connected to static cooling unit.
A.G. COLLACOTT COLLECTION.

Figure 17.8 25,000-litre Fullwood Direct Expansion farm vat.
COURTESY FULLWOOD LTD.

Figure 17.9 Early insulated twin compartment farm collection tanker. Capacity 8,000 litres.
BILL CORNWALL COLLECTION.

Even in the worst weather conditions, the farm road would likely be passable to a tractor or four-wheeled-drive vehicle. It was sensible to advise farmers to have on the farm a suitable container that, carried on a

Figure 17.10 Early treble compartment articulated farm collection tanker. Capacity 11,000 litres.
BILL CORNWALL COLLECTION.

Figure 17.11 Milk being transferred by vacuum from farm collection tanker to 20,000-litre insulated tanker for distant delivery to urban market, both owned by MMB.
BILL CORNWALL COLLECTION.

tractor or trailer, can be taken to the tanker at the roadside.

If the tanker is delayed for more than one milking, an emergency container could also be used to store

Figure 17.12 Drawbar unit, for hauling milk over long distances. Capacity 18,000 litres.
Bill Cornwall collection.

Figure 17.13 Standard vacuum loading tanker. Capacity 9,000 litres.
A.G. Collacott collection.

milk awaiting collection; permitted only in a genuine emergency.

Emergency containers provided an inexpensive insurance against loss of milk. There were a variety of

Figure 17.14 Modern 30,000-litre farm collection tanker.
COURTESY TURNERS OF SOHAM.

containers recommended by the Milk Board, including the more expensive rigid tanks or low-cost plastic bags with inlet/outlet for tanker hose, designed to be carried on suitably prepared farm trailers so as to avoid sharp objects puncturing the bag. These bags could only be used once but were cheap to buy and occupied a minimum of storage space. Emergency containers were available through the Milk Board at a favourable price.

Rigid containers varied in size from 100 gallons up to 375 gallons. Some were made to be attached to the three-point linkage on a tractor and some carried on a buck rake or link box.

All the rigid emergency containers could be used to carry water or other non-toxic liquids and easily cleaned after use by regular iodophor dairy cleaning chemicals. Other suitable uses for these rigid containers was the transporting of milk from mobile milking bails in fields to a refrigerated vat at the farm building.

18

The Task Force

THE milk industry did not escape the industrial action that spread across the country during the late 1960s and 1970s. One of the first strikes to affect producers and the daily farm collection of milk was at the Board's own creamery at Harford Bridge, Norwich in September 1969. Milk collection drivers, at very short notice, withdrew their labour over pay. The Milk Board's head office received a telephone call from the creamery that there would be no milk collection from around 200 churn and bulk farms in Norfolk that day and until a settlement was reached.

The driver's union was well-aware of the importance of moving such a perishable commodity as milk and the strong position they held in demanding a substantial increase in wages.

They were also aware that a prolonged strike would have a devastating impact on the income of milk producers, as well as on the daily supply of milk to consumers. The cost of transporting milk from farm to processors was a direct cost to producers, and large pay award would have a serious effect on farm-gate price across the country.

Strenuous efforts to bring about a quick settlement

failed. Doing nothing was not an option. The drivers' representatives clearly underestimated the Milk Board's resilience and staff's loyalty to producers and delivery of milk to householders, hotels and catering establishments.

The Milk Board sent out a message to staff in possession of a driver's licence for volunteers to drive tankers and churn collection trucks (a HGV licence was not required in those days).

There was an enormous response from all ranks, mostly marketing staff and regional office workers and both male and female. Most had never driven heavy goods vehicles before. Fortunately, marketing officers were fully conversant with bulk collection. Volunteers came from various parts of the country. Bed and breakfast accommodation was hastily arranged. A call went out to members of Norfolk Farmers Union for navigators for those not knowing the area.

Fortunately, late summer and early autumn is a low milk production period in Norfolk, so most bulk consigning producers were able to hold three or four milkings. Many churn producers had a few extra churns available to enable them to store milk without change to their usual milking times. Milk Board supply officers quickly arranged for milk to be brought into the creamery in bulk from other parts of the country to meet immediate requirements to ensure that daily deliveries to the doorstep were not disrupted.

By early evening, volunteers with hastily packed overnight bags began to arrive from across the country. By midnight, all collection vehicles were in operation. The creamery stayed open all night to receive milk. Other volunteers delivered bottled milk out to distributors over a wide area and trucked in goods so as to keep the creamery working.

One volunteer driver not conversant with the local lingo was puzzled when delivering bottled milk to a Milk Board distribution depot. He was prevented from entry by a member of a picket who said 'Don't you know there is a dispoot about here?' He soon realised folk dew things diffrunt in Norfolk.

Within a few days, the milk collection and distribution was back to normal. The strike lasted 74 days and caused considerable disruption in the national bulk conversion scheme. But by not giving in to unreasonable demands, many thousands of pounds of producers' money was saved. The work of volunteers meant not a single farmer lost any milk.

The Norwich strike was a wake-up call for all and highlighted the Milk Board's vulnerability to industrial action or extreme weather conditions. It was decided to compile a permanent list of volunteer drivers to be available at short notice to go to any part of the country. This became known as the Task Force, about 50 in number, made up of marketing officers, line managers and office staff, male and female.

Shortly after the Norwich strike, heavy goods vehicle driver licensing came into force. Thus volunteers were obliged to take part in heavy goods vehicle driver training and pass the HGV driving test.

There were a few female volunteer drivers; one girl in particular who was a little over five foot tall who sailed through an articulated vehicle test and regularly drove a 4,500-gallon (20,000-litre) tanker over long distances during subsequent industrial action.

The dairy industry, like many other industries up and down the country, did not escape sporadic strike action. Added to this was occasional extreme weather conditions, during which the Task Force volunteers played a useful role in saving producers loss of milk, as well as keeping the nation supplied with their daily delivery of fresh milk.

19

The Milk Marketing Board's Commercial Enterprises

IN addition to providing a secure market for their milk and stability in the industry, the Board did enormous amount of good for our milk producers and the industry as a whole.

The Board provided much on-farm help to producers at a reasonable cost. In 1943, the Board became responsible for milk recording, hitherto carried out by county committees. National Milk Recording (NMR) provided a valuable service to dairy farmers, especially pedigree breeders whose societies accepted the NMR as a reasonable record of each cow's annual milk and butterfat production.

As a further aid to dairy farmers, in the 1950s, the Board set up a Private Milk Recording service (PMR); a less expensive service than NMR. They also set up a milking machine testing service as well as a consultancy service providing expert advice on herd management and nutrition.

Before World War II, the Board began to set up its own creameries, gaining valuable knowledge of the cost of manufacturing dairy products; a valuable aid when negotiating the price of milk sold to private companies for manufacture, together with milk retail

Figure 19.1 Dairy Crest, Torrington, Devon (now closed).
A.G. COLLACOTT COLLECTION.

outlets later to come under the control of the wholly producer-owned, highly successful company Dairy Crest.

The Milk Board's commercial enterprises brought huge financial reward to the milk producers of England and Wales, generating huge sums sufficient to cover the entire cost of running the Milk Marketing Board and the cost of transporting milk from dairy farmers to buyers, as well as increasing the Milk Fund to the benefit of all producers.

Sadly when their wholly owned organisation that served them so well for more than 60 years was wound up in 1994 by the government, neither the milk producers nor their leaders made any meaningful effort to save it, in the mistaken belief they would be better-off in a free market economy. The consequences to the industry was devastating. Thousands gave up milk production altogether.

Figure 19.2 Continuous butter-making plant, Dairy Crest, Torrington.
A.G. COLLACOTT COLLECTION.

Figure 19.3 Butter churning, Dairy Crest, Torrington.
A.G. COLLACOTT COLLECTION.

Figure 19.4 Packaging skim powder, Dairy Crest, Torrington.
A.G. COLLACOTT COLLECTION.

20

Artificial Insemination (AI)

BY far the greatest enterprise set up by the Board was artificial insemination of beef and dairy cattle.

The history of artificial insemination of animals is interesting. According to some papers, the earliest reports of insemination took place as far back as 1322, when an Arab chief used artificial methods for the successful insemination of a prize mare; apparently using semen collected from the sheath of a stallion belonging to an enemy chieftain. This was also a practice subject to research and experiments in 1890 in the horse-breeding industry in Germany, France, Holland and Denmark.

The first artificial insemination of cattle cooperative was established in Denmark in 1933, followed by similar establishments in the United States. After the Second World War, Doctors Edwards and Polson carried out experimental work on artificial insemination of cattle at Cambridge. Shortly after, various organisations introduced the service to farmers across England and Wales.

It was under the Agricultural Insemination Act of 1946 that the Milk Marketing Board was asked by the government to provide artificial insemination of cattle to farmers across England and Wales.

Realising the enormity and benefits of artificial insemination of cattle, a small group of farmers in 1943 set up North Suffolk Cattle Breeders at Benacre, near Lowestoft, headed by Geoffrey Smith, a newly qualified veterinary surgeon from Loddon in Norfolk, providing the service to farmers in north Suffolk and south Norfolk. The Milk Marketing Board saw the value of this service and took over the enterprise, operating from a purpose-built cattle breeding centre on Ringsfield Road, Beccles with Geoffrey Smith in charge.

Exercising the government's invitation to provide the service across England and Wales, the Board rapidly set up 23 main centres and 63 sub-centres. Each main centre had a team of technicians, a bull stud, mainly British Friesian, as well as Ayrshire, Dairy Shorthorn, Jersey, Guernsey, Hereford, Aberdeen Angus, Red Poll, north and south Devon breeds from which semen was collected, processed and distributed to sub-centres, each under the control of a senior technician providing the service to outlying areas. So popular was the service that by 1982, the MMB plus four privately run centres – two Ministry of Agriculture at Reading and Ruthin in north Wales – were carrying out nearly two million inseminations each year.

Before approval for artificial insemination use, a bull's pedigree would be thoroughly researched and, where possible, the bull's progeny would be inspected by expert breeders. The bull was also subject to extensive health tests by one of the Board's veterinarians.

The actual technique of inseminating a cow was quite simple. In the early days of commercial artificial insemination, a trained technician would locate the uterus through the rectum and then insert 1cc of diluted raw semen into the body of the uterus using a glass pipette and syringe.

A bull in good condition would produce around 4cc of spermatozoa, up to 500 times more than that required to enable a cow to conceive. A skilled technician would evaluate the sample, paying particular attention for abnormalities (very rare), density and movement.

The sample would then be diluted in a solution of egg yolk and citrate sulphur animide, protecting sperm cells from shock during cooling to 4.5°C. The sample was then re-examined and approved for use that day. Semen on the technician's round was stored in a thermos flask containing an ice pack. A technician would visit 20 or 30 farms each day, inseminating 30 or 40 cows with a success rate of 68–70 per cent. To receive the service, the farmer paid £1 to become a member and in the early days 17 shillings and six pence (87.5 new pence) for each animal, including two free services should it be necessary within three months of the last service. This fee gradually rose to £1.25 for the bull of the day and remained at this level for many years.

Semen can now be stored for years using modern techniques. Glycerol is added to the sperm diluent, which allows sperm to withstand freezing without losing fertility when thawed. Diluted semen placed in straws, stored and frozen using liquid nitrogen, enabled a farmer to nominate a bull to suit his particular breeding programme for many years ahead.

In the early days, there was much scepticism about the method of getting cows in calf by artificial means.

Among a variety of wild claims were that calves would be born with two heads, six legs, undershot jaws, overshot jaws, some even suggested artificial insemination progeny were unlikely to live long. One well-known dairy farmer near Beccles in Suffolk said he would never use artificial insemination because he did not wish to deprive cows the pleasure of natural mating. Cattle-dealers who made money buying and selling on bulls of dubious pedigree as well as hiring out bulls to small farmers were quite happy to throw fuel on the fire of dissent. In fact, the presence of transmittable diseases through natural mating using unhealthy bulls fell quite rapidly.

Figure 20.1 In the early years of artificial insemination there were many strange stories...
A.G. COLLACOTT COLLECTION.

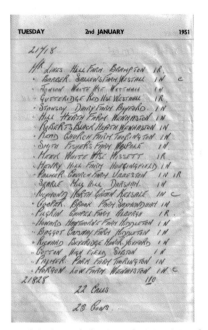

Figure 20.2 Typical day's work for AI technician (N = first service, R = repeat service, C = cash payment).

A.G. Collacott collection.

Wait till he gets here then tell him Master's let cow out

Figure 20.3 AI technician struggling through snow to artificially inseminate the farmer's cow.

A.G Collacott collection.

Technicians were often referred to as the 'bull with a bowler hat'. Even amateur poets got in on the act. The following poems are most certainly worth sharing. Sadly the poets of most are not known.

Understand a Bull

I'm Billy the bull from the Centre
My semen is sent far and wide
And when there is a rush
I'll be put in a crush
To perform on a moth eaten hide

When I was young I had freedom
My wives I would woo like of yor
But now I must thrust
And do it or bust
As science demands more and more

My father was very famous
Of his fifty daughters in milk he would boast
But if he knew his son
Had three thousand and one
He would arise from the dead as a ghost

Now, when I am old and so past it
And in the crush I am no longer required
I pray that I can
Without the aid of a man
Have one cow before I'm retired

My fame it is spread over England
My offspring are bought for top price
And if you will pay the fee
You can still have it from me
And there's cartons and cartons on ice

No Bull

My mother, her name is just Daisy
She's not had a bull by the horn
She swears to me that she is still a virgin
Then how the hell was I born

I questioned my dear old Great Granny
She say my father just came in a straw
He was brought by the veterinary surgeon
Which for my mum was a terrible bore

Gran say it's sad to be one of 10,000
Much better to be one of the few
And she says 'oh my dear little heifer
I expect it will happen to you'

Are the days of cows love and passion
All gone so we can't have a thrill
Is a bull really out of fashion
And replaced by a technician called Bill

Gran says let us all get together
Go on strike and not get in calf
Tell them to take the old straws and just stuff them
Then we will all get the last laugh

Now farmer Brown just you try it
Give your missus a straw in her bed
And we are sure if she is a real woman
She'll bash you until you are dead

I Was Not Brought About by a Bull

Now I've just given birth to a heifer
And pride and of milk I am full
But sad to relate, this lactical state
You see I was not brought about by a bull
I've never been naughty, I swear it
In spite of this calf that I have borne
By farmer Brown's tractor, I'm virgo intacta
And I've never had a bull by the horn

How dreary the fields and meadows
Sheep pens are gloomy and grey
For the day's bit of fun in the years dreary run
Has by science been taken away?

Now they say that the farm is a business
In which we must pull our weight
And I'd pull and I'd pull for a strong built bull
But this phoney arrangement I hate

Now it mustn't be thought that I'm jealous
There are things that a cow shouldn't say
But these Land Army tarts that handle our parts
Still get it the old fashion way
*(By kind permission of Tom Brown, from a song when at
Brooksby Agriculture College, mid-1960s)*

No, No More Artificial Insemination
The missus is really disgusted
She asks us why we are not in calf
When we tell her the natural reasons
All that she does is laugh

We tell her we really can't stand it
This chap with his 'straws' is too bad
It is high time she went to an auction
And bought us a four legged lad

We don't mind if he is heavy
We really don't mind if he stinks
We're sure we would really prefer it
Whatever the human race thinks

The missus she gave us a birthday
The missus she gave us a treat
She bought us the bull from the Centre
Whose semen had come here each week
And you know, he is really good looking
Much better than we all thought before
And he really does it quite nicely
Much better than the old frozen straw

125

He jumped us all with such vigour
That was the end of all the thrill
And now, Oh just look at my figure
I wish I had been on the pill

In 1950, technicians were paid £5 for a six-day week. No overtime was paid, nor were there any perks. The service was increasing rapidly, which often meant working as late as 8pm or 9pm. Working on one's day off did not mean extra pay but a day off in lieu when the workload permitted. Each technician's conception rate was constantly monitored. Poor success rate meant an immediate period of retraining and, if still unsatisfactory, dismissal.

Farmers quickly realised the benefits of artificial insemination as a means of speeding up genetic improvements from bulls that many could not afford. The rapid growth of the service was quite remarkable. By 1950 there were, in addition to four privately run centres, 800 Milk Marking Board technicians working across England and Wales. The Board's bull stud expanded to nearly 1,000 and by the end of 1958 accounted for ten million cattle bred by the service from 15 different breeds. Other benefits allowed producers to build up closed herds, thus providing disease protection as well as avoiding the necessity of keeping a dangerous bull. It is estimated that more than 90 per cent of dairy cattle in the United Kingdom are now artificially inseminated.

For Geoffrey Smith and his team at Beccles, 22 November 1951 was a very sad day. There were sporadic outbreaks of foot and mouth disease in Suffolk and the artificial insemination service was suspended in certain areas by the Ministry of Agriculture.

Technicians helping with the bull stud were not allowed on farms. The place was awash with disinfectant and virtually sealed off from the outside world. Even trucks delivering feed were not allowed on the premises; these and other deliveries were left on a scrupulously cleansed trailer at the roadside.

Early one morning, a technician noticed an Ayrshire bull that had been in an isolation box for several weeks showing foot and mouth symptoms, which subsequently proved positive.

A total of 37 bulls and two cows were slaughtered. A huge pit was dug close to the centre buildings where they were shot and their carcases pushed into it, then covered with quicklime followed by several feet of earth. It was a sad day and a dreadful loss of some of the country's finest breeding stock. Veterinarians were adamant that the disease was brought into the Beccles Centre by birds, but there was no way of proving that this was the case.

Outbreaks of the disease continued in Suffolk and Norfolk into the New Year with the tragic loss of a few well-known pedigree herds. Staff were laid off or sent to other artificial insemination centres. By mid-January 1952, the foot and mouth outbreak in East Anglia abated and service began to get back to normal. Until a replacement bull stud could be built up, the centre and its sub-centres at Aylsham, Wymondham and Otley in Suffolk were supplied with semen from artificial insemination centres across the county.

Abolition of the Milk Marketing Board in 1994 resulted in Genus Limited becoming responsible for the breeding and production services formerly provided by the MMB in England and Wales. Genus also provides training for farmers and staff to inseminate their own cattle and enable them to purchase frozen semen from bulls of their choosing and storage on their farms.

Technology has moved on from the pioneering days of raw semen, when much of it was wasted. Semen is now placed in straws and can be frozen and stored for a long period of time, thus maintaining breed blood-lines long after the donating bull has died.

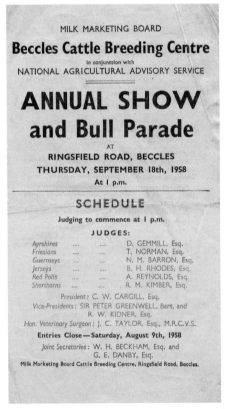

Figure 20.4 Popular annual bull parade. MMB Beccles Cattle Breeding Centre.

A.G. COLLACOTT COLLECTION.

21

Rise in Milk Production

ARTIFICIAL insemination of dairy cattle across Europe rapidly enhanced individual cow yields. The 1960s and 1970s saw a vast improvement in milking techniques. Specialised equipment and bulk handling of milk on dairy farms enabled large herds to be milked efficiently and quicker. New specialised dairy units backed by handsome government cash grants resulted in massive overproduction and huge stocks of cheese and butter – which became known as 'butter mountains' – at vast cost to the public. Thus, in 1984 'milk quotas' were introduced for producers in the European Economic Community (EEC) states. It was unfortunate that quota regulations were formed without regard for the large number of exceptional and problem cases that occurred and that caused a great deal of unnecessary distress. Many producers adopted panic measures to avoid having to pay the 'super levy' on milk produced over quota. As it turned out, production was below quota at the end of the first year, dropping from an all-time high of 12,775 million litres in 1980 to 11,211 million litres in the year ending March 1994.

The effects of milk quotas were not only felt by

farmers but also millers, suppliers of dairy equipment, farm buildings and various service agencies.

Hitherto, overproduction meant more milk going into the manufacturing market and a drop in the price paid to farmers for their milk. Quotas had a cash value and could be sold on to the other producers. Many small dairy farmers and those contemplating retirement took advantage of this and gave up milk production altogether. Introduction of bulk collection also contributed to producers going out of milk. At the time of the winding up of the Milk Marketing Boards in 1994, the number of farmers producing milk had dropped to 26,000.

Many milk producers were of the opinion that the MMB was a government body run by civil servants. It was not. It was wholly milk producer-owned; run by milk producers. None of the Board members received a salary. The Board's commercial enterprises not only covered the cost of running it but provided handsome sums to the Milk Fund for payment to all producers.

The system, managed by the MMB, worked well for both producer and consumer. There was stability in the marketplace. Bank managers valued the monthly milk cheque and were happy to loan money for reinvestment. This helped young farmers to get going and many were able to purchase their farms on assurance of a milk contract with the MMB. Orderly marketing prevailed, which kept 'milk miles' to an absolute minimum. There were no farm collection vehicles running over each other's territory. There was no middleman between producer and processor.

At its peak, the annual turnover of the England and Wales Board was in the order of £2,600 million purchasing milk from 160,000 producers.

The work of the wholly producer-owned UK Milk Marketing Boards is probably best summed up by John Empson under the heading *The UK Marketing Boards: A Concise History*, published by the Society of Dairy Technology. In the introduction, John Empson writes: "The Milk Marketing Boards were, by any measure, the greatest commercial enterprise ever launched by British Farmers. Indeed in terms of their marketing activity, the collection and sale of raw milk from farms, they were the largest such organisation in the world. At its peak, the England and Wales Board marketed annually some 13,000 million litres of milk, twice that of its nearest international rivals.'

22

Winding Up of the UK Milk Marketing Boards, 1994

IN response to a question from Mr Tyler in the House of Commons on 27 January 1994, concerning the winding up of the UK Milk Marketing Boards, the then Minister of Agriculture, Gillian Shephard said: 'He should be perfectly reassured that the interest of 29,000 milk producers are absolute top priority, that interests of 50 million consumers also have been taken into account.'

Despite a provision in the Milk Marketing Scheme to protect the consumers, in 1994, arguably to comply with EEC Common Agriculture Policy (CAP) and our own Office of Fair Trading (OFT), our government decided to wind up the UK Milk Marketing Boards and regulatory pricing on the grounds that they were a monopoly organisation contrary to consumer interests, even though producers throughout the lifetime of the Milk Boards were able to retail their milk to consumers and as much farm bottle milk as they wished to dairy companies.

While provisions were put in place via the Office of Fair Trading to protect consumer interests, nothing was put in place to protect milk producers from the enormous power of the rapidly growing supermarkets.

Figure 22.1 The end of the Milk Marketing Board in 1994. COURTESY A.G. COLLACOTT COLLECTION.

Producers were warned by the Milk Marketing Boards that free market incentives being bandied about by dairy companies would benefit them only in the short term. Sadly, neither dairy farmers nor their leaders made any meaningful effort to save their own organisation that had served them well for more than 60 years.

In the final year of the Milk Marketing Board of England and Wales, producers were paid 24.47 pence per litre (ppl). By 2006, it had dropped by 26.85 per cent to 17.9ppl against rising cost of production, while the price to the consumer increased by 20.40 per cent from 42.3ppl to 50.9ppl. In contrast, retailers, of which supermarkets had by far the greatest share, took full advantage of the free market economy and weak bargaining position of dairy farmers.

In 1995, producers received 57.91 per cent of a litre of milk sold on the liquid market. By 2015 it dropped to 40.86 per cent, while the processor/retailers share rose to 59.14 per cent.

Price to the consumer has gone up by 44.87 per cent. (Figures based on a leading supermarket pre-loss leader campaign retail price of £1.39 for four pints, 61.17ppl.) These figures do not take into account the huge processors' benefits from cream by-product or from semi-skimmed and skimmed milk production.

Thus, the winding up of the Milk Marketing Boards by the government has not achieved what was intended.

It is difficult to have sympathy with dairy farmers. Without the support of cooperative marketing, they have allowed themselves to fall back to the 'dreadful days' before the Milk Marketing Boards were set up in desperation by their forefathers who had the foresight that cooperative marketing was the salvation of the dairy sector. By 2007, 15,000 dairy farmers, more than 50 per cent, gave up milk production. At one point, more than 1,000 a year were going out of milk production, with many in the West Country allegedly close to becoming bankrupt. In some areas, producers were crying out for reinstatement of the Milk Marketing Boards. The industry was in a mess.

Many well-established pedigree herds built up over generations of farming families disappeared. A quarter of a million cows were reported to have been slaughtered. Home production of milk was looking over the precipice of disaster and talk of having to import milk was rife.

The plight of our dairy farmers was brought to the attention of the nation by a brilliant public awareness campaign initiated in 2005 by the National Federation of Women's Institutes (WI), which led to a 72,000-signature petition backed by the Farmers' Union of Wales for Action, together with the Small Farmers Association. It requested a full-scale government investigation into the practices of the dairy

Figure 22.2 Fresh milk, from Europe.
A.G. Collacott collection.

sector and for a watchdog body to ensure all parties in the food chain received fair prices.

It is reasonable to credit the WI's Great Milk Campaign, in which many meetings were organised across England and Wales, with stirring some of the larger supermarkets into realising that to maintain a supply of fresh milk from British farms, producers had to be paid a fair price for their milk. They did this by entering into direct stringent contracts with selected producers, seemingly regardless to geographical location of the producer's farm.

Keeping these supplies separate meant collection tankers running over each other's collection fields, as had been the case before the setting up of the Milk Marketing Board of England and Wales, thus increasing milk miles, carbon emissions and, finally, the end of orderly marketing.

23

A Reminder of what the Milk Marketing Board of England and Wales Did for Milk Producers

MINDFUL of the state of dairy farming before the setting up of the Milk Marketing Board of England and Wales, it is worth bringing together some of what the MMB did for our producers:

- A binding milk contract for life, bringing with it stability in the milk producing industry.
- Undertook to purchase and sell all the milk a producer wished to produce.
- Promoted sales of raw milk to the most remunerative markets.
- Set up its own creameries and retail outlets under the title Dairy Crest.
- Developed new dairy products, such as Cathedral Cheese and Clover Spread, as well as many other products.
- Introduced milk quality payment scheme to encourage the production of good quality milk.
- Introduced on-farm refrigeration of milk and bulk collection from every producer in England and Wales.
- Set up bulk buying groups for farm vats, saving producers thousands of pounds.

- In conjunction with the dairy trade, set up the Liaison Chemist Service to ensure milk was tested at buyers' premise by approved staff and in accordance with joint committee Code of Practice.
- Set up its own fleet of milk collection vehicles.
- Transported milk from farm to buyer in the most economical manner.
- Organised transhipment points in collection milk fields to tranship milk from small farm collection tankers to large articulated tankers and drawbar units for forwarding to distant markets, thus cutting out costly country depots.
- Set up voluntary Task Force to drive milk collection vehicles in the event of industrial action by drivers and to assist during extreme weather conditions.
- Took over from county committees National Milk Recording Service, providing individual cow yields and butterfat content.
- Set up one of the largest artificial insemination services in the world, thus allowing small farmers to have the use of the country's top-class bulls and improve the genetic quality of their herds.
- Provided farm management services, including breeding programme, Private Milk Recording and Milk Machine Testing Service.

In the event of an emergency, MMB marketing staff were available 365 days a year.

All the profits from the Board's commercial enterprises covered the cost of running the MMB and transport costs of getting milk from farm to market. A handsome surplus from these enterprises was added to the farm-gate price paid to producers.

The question remains as to why producers and their leaders failed to fight tooth and nail to save their very own organisation. Did we lack the French farmers' spirit in fighting for the finest wholly producer-owned marketing organisation this country has ever known?

Why did producers not set up a national or regional cooperative marketing organisation as retained by other EEC member states? Around 90 per cent of milk produced in Northern Europe and Southern Ireland is sold to producer-owned cooperatives, why not England and Wales?

24

The Future

OUR dairy farmers no longer have the protection of the Milk Marketing Board, nor government price regulation, which protected both producer and consumer.

Why was there no effort to encourage volunteer cooperative marketing similar to that retained in North Europe and Southern Ireland?

World shortage of milk production in recent years has resulted in an improved price for milk going to the manufacturer. This market has been, and always will be, a highly volatile subject to world production. Already production in Australia and New Zealand is rising against a dropping Asian market, as well as an export ban on dairy products to Russia. Abolition of milk quotas in Europe in 2015 is likely to increase production. Since the winding up of the MMB sales into the liquid market, the farmers' hitherto most stable and profitable market, has fallen from around 55 per cent to 45 per cent. It is my strong belief that the liquid market will gradually return to the most profitable for our milk producing industry.

With modern microfiltration methods, the life span between farm and consumer is much greater, thus

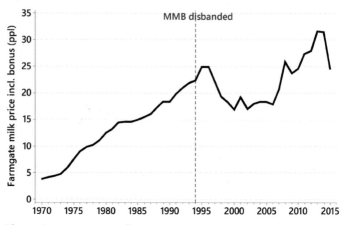

Figure 24.1 Farmgate milk prices 1970–2015.
REDRAWN FROM DEFRA DATA.

opening up our London and southeast liquid market to mainland dairy farmers' cooperatives. It is not unreasonable to assume that the owners of the huge new processing plant in Buckinghamshire already have this in mind. Time will tell.

In the short term, I believe the price paid to producers is likely to continue to fall, especially where processors need to unload stored manufactured products, possibly bottoming out at 25ppl or less.

To lessen further reduction in price paid to farmers, less milk should be subject to world markets. Producers cannot expect buyers to buy what they cannot sell, which in extreme circumstance could lead to some producers not able to find a market for their milk at all.

The UK has a £1.25 billion deficit in dairy products and our producers need to cut production costs to tap into this home market and ward off raw milk and milk products imports. Sadly, the industry is in a

Figure 24.2 Milk production in the 1920s.

mess, and not dissimilar to the situation it found itself in during the 1920s.

Something urgent needs to be done before British milk disappears from retailers' shelves. Producers cannot continue to sell their milk below production costs. In my view, the government has a duty to save our milk producing industry. They must bring dairy farmers' representatives and the dairy trade together and set an annual farm-gate price for milk sold on the liquid market that would enable farmers to produce milk at a price to provide a reasonable living and investment in equipment. Let us, for instance, assume the nation's requirement matched that sold on the manufacturing market and the annual review liquid selling price was fixed at 36ppl, the producer would receive 36ppl for half of his/her supply and half, say 18ppl, for milk sold for manufacture, an overall price paid to the farmer would be 27ppl. Such a pricing system would deter individual producers from the overproduction that currently exists, and bring production more in line with the annual review price. All milk buyers should be obliged to pay for milk supplied to them to a central fund and producers paid by the 20th day of the month following delivery. This system could be administered by a joint government/producer body. Milk buyers would be debited the agreed annual review price for all milk supplied to them less an appropriate negotiated rebate relative to the particular product manufactured similar to that operated by the MMB for many years.

Going it alone, which many producers yearned for, has not worked. Since the winding up of the Milk Boards and deregulation, nearly 20,000 UK milk producers have gone out of milk production. Today there are less than 10,000 and I fear this trend is likely to continue.

Producers will need to look at production and buying costs, there may be some benefit in looking at producer-owned Southern Ireland dairy farmer cooperatives, where milk is purchased, manufactured and sold on their home market as well as abroad. Other significant services are provided to milk producers, such as bulk buying of cattle food and dairy equipment, which are then sold on to their members at competitive prices.

Setting up 1,000-cow-plus specialist units may be a short-term answer. It was not so many years ago when a producer could earn a reasonable living from

15 cows. So will 1,000-cow units become 2,000 and more? Such intensive means of production may well raise welfare issues for cows who are natural grazers of grass spending the best part of their lives on concrete. I fear that herd size will not cut any ice with their milk buyers. Buyers have a duty to their shareholders and will, if the product is right, go to the cheapest supplier, even if they are to be found in mainland Europe.

Other than run of the mill expenditure, there may be other ways worth exploring to cut costs. Milk production has high all-year-round power and water costs. Installation of solar systems might offset the cost of electricity along with the storing up of roof water for washing down plant and yard areas. Surplus electricity could be sold back to the National Grid or even used to dry slurry for the garden centre trade or arable farmers.

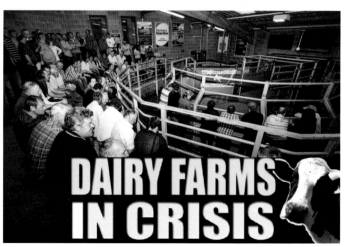

Figure 24.3 Milk production in 2015.
Courtesy Shropshire Star.

Acknowledgements

I am grateful to Professor Tim Sparks, Coventry University, for his kind help and support; my former colleague Bill Cornwall; Suzy O'Shea at Fullwood; writer and broadcaster Keith Skipper; Nigel Pickover, editor at the *Eastern Daily Press*; Mark Wright, editor at the *Shropshire Star*; Jessica Coleman, Lawrence Lamborn and others for providing valuable images.